Mordecai C. Cooke

Our Reptiles and Batrachians

a plain and easy account of the lizards, snakes, newts, toads, frogs and tortoises

indigenous to Great Britain

Mordecai C. Cooke

Our Reptiles and Batrachians
a plain and easy account of the lizards, snakes, newts, toads, frogs and tortoises indigenous to Great Britain

ISBN/EAN: 9783337388188

Printed in Europe, USA, Canada, Australia, Japan

Cover: Foto ©berggeist007 / pixelio.de

More available books at **www.hansebooks.com**

OUR REPTILES

AND BATRACHIANS

A PLAIN AND EASY ACCOUNT OF THE

LIZARDS, SNAKES, NEWTS, TOADS, FROGS AND TORTOISES

INDIGENOUS TO GREAT BRITAIN

BY

M. C. COOKE, M.A., LL.D., A.L.S.

Author of " Rust, Smut, Mildew, and Mould," "A Plain and Easy Account of British Fungi," "Manual of Structural Botany," "Manual of Botanic Terms," etc., etc.

WITH ORIGINAL FIGURES OF EVERY SPECIES

AND NUMEROUS WOODCUTS.

NEW AND REVISED EDITION.

LONDON

W. H. ALLEN & CO., LIMITED

13, WATERLOO PLACE, S.W.

1893

WYMAN AND SONS, LIMITED, LONDON AND REDHILL.

PREFACE TO THE FIRST EDITION.

It may cause surprise to some who are not amongst my most intimate friends, that my name should be attached to a volume on any other branch of Natural History than one which is in some way associated with the Vegetable Kingdom. To these it may be necessary to explain that I have only returned on this occasion to an "old love," long deserted for the fascinating charms of "moulds and mildews." In more youthful days I bird's-nested, caught butterflies and worried reptiles, with all the pertinacity of youth, and amongst all these pursuits acquired a taste for the study of our native birds, reptiles, and insects, which led to my first lessons in the classificatory sciences. From these, it is true, I diverged in

later years, and almost confined myself to plants;
but in a foolish moment, perhaps, the old hankering
to have a word or two with one's "first love" has
come over me, and resulted in this humble account
of "Our Reptiles." I make no pretensions to the
production of anything more than a popular volume
on a rather unpopular subject, to the espousal of the
cause of a much-abused and scandalised class; and
if I only aid in recovering their character from a
little of the obloquy which attaches to them, I shall
not regret the venture. The man of science, if he
seeks for that which is novel or abstruse, had
better close the book and go no further. I do not
presume to have written anything for him; but for
those who know little or nothing of the subject, I
hope that herein may be found a useful introduction,
and a trustworthy guide. The order of the chapters
is not precisely that of the classification of the animals
described, but in the Appendix a Systematic Arrangement has been pursued. In conclusion, I
acknowledge with pleasure the kind and courteous
assistance rendered to me by Dr. Albert Gunther
and Dr. J. E. Gray, of the British Museum, in pro-

curing figures to illustrate the work, as well as for
hints and suggestions in its prosecution.   To those
who honour it with a perusal, I now commend it
with "all its imperfections on its head."

UPPER HOLLOWAY.   M. C. C.

PREFACE TO THE REVISED EDITION.

THE original edition having been for a long time exhausted, the publishers requested me to revise and make such additions as were necessary to bring this little volume up to date. In doing this I have discovered that very little was necessary beyond a more definite recognition of the differences which subsist between the true Reptiles and the Batrachians, in accordance with the views of the most celebrated zoologists of the day. I have permitted the chapter on Gray's Banded Newt to remain, and also the plate, although under the impression that the species has been introduced into the British list by error. The flattering reception which was accorded to the first edition by the most competent authorities in this country, encourages me to hope that the present will share the good fortune of its predecessor, and maintain its position in Popular Natural History.

April, 1893.   M. C. COOKE.

CONTENTS.

PLATE I

1. Sand Lizard

2. Viviparous Lizard

INDIAN SNAKE-CHARMER.

(*From a Sketch by a Native Artist.*)

REPTILES AND SNAKE-STONES.

REPTILES, in zoology, as interpreted forty years ago, constituted a *Class* of vertebrate animals (that is animals with a backbone) intermediate between birds and fishes, having a greater affinity with the latter than the former. They were generally described in scientific works as having cold blood, being the possessors of a heart with sometimes one and sometimes two auricles, but with only one ventricle; so that, at most, there are but three chambers in their

B

hearts instead of four. They were still further characterised as oviparous, breathing by lungs, or partly by lungs and partly by means of gills; thus combining one of the elements of fish-life with those of higher organisms. Finally, their scientific portrait was completed by the announcement that the body is covered with shelly plates (as in the tortoises, &c.), or with scales (as in the snakes), or with a soft naked skin (as in the toads and frogs).

This was the orthodox definition of what constituted a Reptile, but not the best definition, as Mr. Edward Newman afterwards showed in some important remarks on the classification of these animals.

"The epidermis," he writes, "or outer skin of quadrupeds, is clothed with hair, of birds with feathers, of fishes with scales, but in reptiles it is uncovered, perfectly naked. The processes, whether described as *squamæ* (in Latin), or *écailles* (in French), are projections, folds, or rugosities of the under-skin; and are not deciduous like hairs, feathers, and scales, but are as permanent and durable as the bones themselves. This may be seen when the slough of a snake is found. This slough is continuous, and contains a faithful mould of each of these processes; it is a very beautiful and very instructive object. The tortoise exhibits the peculiarity of an *articulated skin*, the articulation being clearly discernible in the living animal, but becoming more conspicuous after death, when dehiscence takes place and the plates fall off, perfectly detached from each other."* Let anyone examine the cast-off skin, as it is called, of a viper or snake, and he will find it to be a thin delicate cuticle which had covered all the projections and inequalities of the true skin, containing a little pouch for each of the "scales" (falsely so called) into each of which a projection had extended. True scales are easily rubbed off from the skin in fish, but there is no rubbing them from that of a snake; they are permanent projections with a scale-like form. If there is any value in words, then "scale" cannot be applied at the same time to the deciduous, flat, horny plates of fish, and the flat, depressed, but persistent irregularities of the skin in reptiles. However, with this reservation, we shall proceed to call them "scales" in deference to custom, and the collective wisdom of those more learned in reptiles than ourselves.

This was the position when the first edition of this work was published, and the Amphibia, or Batrachians, were regarded as an order of the class Reptilia. This position cannot, however, be any longer maintained, and hence we have no other

* *The Zoologist*, p. 8450.

alternative than to separate the two distinct groups
of organisms here treated of, into two independent
classes, the REPTILIA, which include the tortoises,
snakes, and lizards; and the AMPHIBIA, or Batrachians, to which belong the toads, frogs, and newts.
It is hardly necessary to state that *Reptilia* is
derived from the Latin word *repto*, "I crawl," in
allusion to the crawling habit of many of its members; and that *Amphibia* is of Greek origin, and
indicates the two-fold manner of life, at first aquatic,
and afterwards terrestrial, of the various species. It
has been pointed out by Professor Huxley that the
Reptilia are more nearly allied to the Birds, and the
Amphibia to Fishes.

Out of respect to the "red-tape" of science, we
must attempt a definition of these two classes, as
now recognized, and practically adopted by Jenyns,
in 1835.

REPTILIA.—Oviparous animals, breathing by lungs;
body covered with scales or shelly plates; impreg-
nation by sexual union; the young undergoing no
metamorphosis.

AMPHIBIA, or BATRACHIANS.—Oviparous animals,
breathing first by gills, and then by lungs; body
covered with a soft naked skin; feet without claws;
eggs impregnated after exclusion; the young undergoing metamorphosis.

There are also anatomical differences, which need

not trouble us to investigate, the above broad features of distinction being sufficient for general purposes. Huxley appears to rely chiefly upon this distinction, that *Reptilia* are "devoid through life of any apparatus for breathing the air dissolved in water," and the *Amphibia* are "always provided at a certain period, if not throughout life, with branchiæ."

The heart and blood are two important points of difference between reptiles and batrachians and the higher vertebrate animals. These former are all cold-blooded. They possess a heart, it is true; but, as compared with higher organisms, an imperfect one, inasmuch as it has but one ventricle; the result of this is that respiration is imperfect, and as respiration gives heat to the blood, which in turn sustains the heat of the body, it follows necessarily from their organization that the temperature should be very low. Let a frog leap upon your hand, or take a newt between your fingers, and the chilly, smooth, apparently slimy appeal to the sense of touch will carry conviction far swifter than argument.

In being oviparous, or producing their young from an egg, these creatures agree with birds as well as fishes; but in some instances the outer covering of the egg, then only a thin membrane, and not a hard shell, is broken at the moment it issues into the world, and the lively young escape, in all respects

miniature representatives of their parents. In the
latter case the term "ovo-viviparous" is generally
applied, and in this sense we may, perhaps, have
occasion to use it. The features of a double life,
wherein one portion is spent in water, breathing by
means of gills, and the other on land, respiring with
lungs, will be illustrated when treating of toads and
frogs and other *amphibians*, so that it will be un-
necessary to enter upon the subject here.

For the purposes, not only of classification, but of
orderly description, the Reptilia are naturally divi-
sible into three orders,—of which the first are the
Chelonians or Tortoises and Turtles, the second the
Saurians or Lizards, and the third the Ophidians or
Snakes. The first, or Chelonians, are scarcely represented in Great Britain at all. The few turtles,
which have been borne in times past upon the waves
that wash our shores, and cast relentlessly upon our
coast, had really no business there, and only came as
occasional, distinguished, and probably involuntary
visitors. Under these circumstances we have given
them a place at the end of the volume, although,
according to rigid science, they should have been at
the beginning.

The Saurians or Lizards are represented in Britain
by four species, three having visible legs, and the
fourth snake-like in form. That the Sand Lizard
and the Viviparous Lizard, as well as the Slow-worm,

are true natives no one will doubt; but whether the
Green Lizard deserves a place in our Fauna is a
more open question, and is again referred to hereafter. The great Gavial of the Ganges, sometimes
nearly eighteen feet long, the Crocodile of the Nile,
the Alligator of North America, and the Caymans of
the South, are the giants of this order; but of these
we have, happily, no representative. In neither of
the two orders named do any of its members possess
poison-bags or venomous fangs, though we happen
to know that it is a firmly-rooted opinion in India
that there is one or more species of lizard capable of
causing death by a wound, rendered mortal either
by a virulent saliva, or some other means. Such a
lizard, however, is entirely unknown to scientific
men, and by them the *Bis-cobra* is believed to be
only a phantom of " the heat-oppressed brain." A
surgeon, for many years on service in India, tells
us that he knew of an instance of a man descending
a well being bitten by such a lizard, that he was
drawn up and indicated the position of the reptile;
that a second man descended and killed the bis-
cobra, which was afterwards preserved in spirits at
the barrack-hospital for many years, and, finally,
that the man who was bitten died in consequence in
a few hours.

The third order includes all the snakes, from the
monstrous *Boa constrictor* and the dreadful Rattle-

snake and Cobra to the Ringed Snake and Viper of
our own islands. Of these we possess but three, two
of which belong to the harmless snakes and the
third to the venomous snakes. There are many
others known on the Continent of Europe which do
not occur on this side the Channel, and in Ireland
those of Britain are also unknown.

We have now indicated the primary groups which
include the nine species found either as true natives,
or naturalized, or as occasional visitors to the British
Isles. Of these, two belong to the Chelonians or
Tortoises, four to the Saurians or Lizards, and three
to the Ophidians or Snakes. Only one of these
nine is capable of inflicting serious injury by means
of its venomous fangs, although, amongst Batrachians, the toad secretes an acrid fluid beneath its
skin, to which allusion will hereafter be made.

" Frogs and toads are found on the Shetlands,
whilst *Vipera berus*, the most northern snake, is
already scarce in the north of Scotland. *Rana
temporaria* is met with in the Alps, round lakes,
near the region of eternal snow, which are nine
months covered with ice ; whilst *Vipera berus*
reaches only to the height of 5,000 feet in the
Alps, and of 7,000 in the Pyrenees. A triton or a
frog, being frozen in water, will awake to its former
life if the water is gradually thawed ; I found myself that even the eggs of *Rana temporaria*, frozen

in ice during seven hours, suffered no harm by it, and afterwards were develope 1. A snake can only endure a much less degree of cold; even in the cold nights of summer it falls into the state of lethargy; it awakes late in the spring, when some frogs and tritons have already finished their propagation; it retires early into its recess in harvest, whilst still the evenings resound with the vigorous croaking of the tree-frogs and the bell-like clamour of *Alytes obstetricans.* Our European snakes die generally in captivity during the winter, partly from want of food, partly by the cold nights. The eggs of our oviparous species are deposited during the hottest part of the year, requiring a high temperature for development. Further, though some accounts of Batrachians enclosed in cavities of the earth or trees may be exaggerated, the fact is stated by men whose knowledge and truth are beyond all doubt, that such animals live many years apparently without the supply of food necessary for preserving the energies of the vital functions."*

In this country, all Reptiles pass the winter in a state of repose, retiring to holes, clefts, or other available places, apparently secure from disturbance, either in company or singly, for a quiet six months' " nap." During this period no food is taken,

* Dr. A. Günther on the Geographical Distribution of Reptiles.—*Proceedings of Zoological Society.*

growth is impeded, the circulation is tardy, respiration is low, and the semblance of death is almost
complete. In the spring the warm sun quickens
their blood, awakens them from their dreams, and
again they crawl or leap into active existence, to
the terror of " unprotected females " and the insect
life on which they prey.

The curious habit they have of changing their
habits, or casting off the outer cuticle of their skins
periodically and then devouring them, offers an
economical suggestion to those who advocate " the
utilization of waste substances," in respect to cast-
off clothes. Indeed, we are not altogether innocent
of devouring our old clothing, but with this difference between ourselves and the reptiles, they appropriate them immediately and in their unchanged
condition, ours suffer mutilation, and manifold
intermediate changes, before they enter our mouths.

Although Reptiles are neither sufficiently numerous nor venomous in our temperate climes to
give the traveller serious alarm, such is far from
being the case in tropical countries. There we consequently hear of snake remedies, snake charms,
and snake charmers, to an almost unlimited extent.
Anyone who has paid attention to the Materia
Medica of hot climates knows how common "snake-
roots" and antidotes to snake poison are in all such
countries. In many instances these substances are

in themselves perfectly useless, and derived their reputation in a great measure from their external resemblance in form to the sinuous or coiled reptile. In many others they are only stimulant or tonic.

The most notable of remedies is the snake stone, not only because of the wonderful powers ascribed to it, but also on account of the belief still entertained, even by many Europeans, of its marvellous curative properties. There is some confusion with regard to it, on account of its numerous imitations. The true snake-stone of the East is undoubtedly a kind of Bezoar or biliary concretion found in the stomach of various animals. Factitious Bezoars are generally either of calcined bone, gypsum, or other absorbent material. The *Zuhr Mohra* or *Zeher Morah*, as it is called in India, is a kind of Bezoar celebrated in Eastern works as a remedy for snake-bites, hydrophobia, &c., and Dr. Ainslie says it is supposed by the Hindoos to possess sovereign virtues as an external application in cases of snake-bites or stings of scorpions; and its various Oriental names imply that it destroys poisons. Dr. Davy, on examining what are called snake-stones in India, found them to be *Bezoars*. The same kind of substance is known in the island of Ceylon under the name of *Pamboo Kaloo*. Berthollet mentions eight kinds of Bezoar, which are chiefly phosphates. These were deemed efficacious not only when taken as medicine, but even when merely carried about the person, so that credulous people would hire them on particular occasions for a *ducat* per day. A single Oriental Bezoar has been known to sell for six thousand livres. The goat Bezoar was found in the fourth stomach of the *Capra ægagrus* of Persia, and was said to be oblong, of the size of a kidney-bean, shining, and of a dark green colour. This was doubtless the most esteemed as a snake-stone.

An account, recently published, of one of these snake-stones, which has great reputation in the island of Corfu, thus describes the manner in which it is employed :—*

When a person is bitten by a poisonous snake, the bite must be opened by a cut of a lancet or razor longways, and the stone applied within twenty-four hours. The stone then attaches itself firmly on the wound, and when it has done its office falls off ; the cure is then complete. The stone must then be thrown into milk ; whereupon it vomits the poison it has absorbed, which remains green upon the top of the milk, and the stone is then again fit for use. This stone has been from time immemorial in the family of Ventura, of Corfu, a house of Italian origin, and is notorious, so that peasants immediately apply for its aid. In a case where two were stung at the same time by serpents, the stone was applied to one, who recovered, but the other, for whom it could not be used, died. It never failed but once, and then it was applied *after* the twenty-four hours. Its colour is so dark as not to be distinguished from black.

In confirmation of the above, another writer adds :—†

* P. M. Colquhoun. in *All the Year Round*, No. 139.
† A. M. Browne, in *Science Gossip*, Vol. i., p. 38.

While in Corfu, where I resided some years, I became slightly acquainted with the gentleman in question, Signor Ventura, of the Strada Reale, Corfu. His family is, as stated, of great antiquity in the island. He does not know exactly when the stone first came into their possession, but conjectures it was brought from India by one of his ancestors. I have myself never seen this remarkable stone, but am fully satisfied as to its efficacy, as I have constantly heard of people being cured by it; in fact, the first thing the Greeks do when bitten by a venomous snake, of which there are several species in Greece, is to apply to Signor Ventura. The stone is then applied, exactly in the manner described above, and the patient in due time is cured.

The instance alluded to where one died while it was being used for another, is of a countryman who was bitten by the viper while cutting myrtle or bay for church decoration. He, as soon as bitten, ran to the town, distant some miles, and arrived when the stone was in use. When it was procured for him, it would not adhere; for it seems this singular stone requires to rest in milk for some time, to vomit, as it were, the poison absorbed. Before it was fit for use again the man died. The stone was broken by a very clever but unscrupulous native physician, who procured it to look at it, as he said, but who broke it in halves, and subjected one half to the most severe tests, totally failing, however, to discover its component parts. The fracture of the stone has slightly impaired its curative power, and, in consequence, I have heard the physician, Dottore ————, railed at in no very measured language by the Greeks.

Sir Emerson Tennent gives an account of the Pamboo Kaloo of Ceylon, which is employed for the same purposes as the above, and the knowledge of such a use he thinks was probably communicated to the Singhalese by the itinerant snake-charmers of the Coromandel coast. "On one occasion," he writes, "in March, 1854, a friend of mine was riding with some other civil officers of the Government along a

jungle path in the vicinity of Bintenne, when they
saw one of two Tamils, who were approaching them,
suddenly dart into the forest and return, holding in
both hands a cobra de capello, which he had seized
by the head and tail. He called to his companion
for assistance to place it in their covered basket;
but in doing this he handled it so inexpertly that it
seized him by the finger and retained its hold for a
few seconds, as if unable to retract its fangs. The
blood flowed, and intense pain appeared to follow
almost immediately; but with all expedition the
friend of the sufferer undid his waistcloth, and took
from it two snake-stones, each of the size of a small
almond, intensely black and highly polished, though
of an extremely light substance. Then he applied
one to each wound inflicted by the teeth of the serpent, to which the stones attached themselves closely,
the blood that oozed from the bites being rapidly
imbibed by the porous texture of the article applied.
The stones adhered tenaciously for three or four
minutes, the wounded man's companion in the mean-
time rubbing his arm downwards from the shoulder
towards the fingers. At length the snake-stones
dropped off of their own accord; the suffering appeared to have subsided; he twisted his fingers till
the joints cracked, and went on his way without con-
cern.' It would appear that Sir Emerson submitted
one of these snake-stones to Professor Faraday for

chemical examination, which resulted in the professor giving his opinion that it was a piece of charred bone which had been filled with blood, perhaps several times, and then carefully charred again. The ash was almost entirely composed of phosphate of lime.

Captain Napier mentions an instance of the efficacy of the stone. One of the soldiers having been bitten by a scorpion, he says, " I applied the stone to the puncture; it adhered immediately, and during the eight minutes that it remained on, the patient by degrees became easier, the pain diminishing gradually from the shoulder downwards until it appeared entirely confined to the immediate vicinity of the wound. I then removed the stone; on putting it into a cup of water, numbers of small air-bubbles rose to the surface. In a short time the man ceased to suffer any inconvenience from the accident."*

Who will deny the evidence of such facts, simply because they cannot understand them? Mr. E. Newman remarks very pertinently on this same question:—" I have often been astonished at the ridicule thrown over facts that we cannot understand. Men of learning who laugh at a phenomenon they have not seen, always remind me of giggling girls who titter when they hear two persons speak

* Gosse's " Romance of Natural History."

any language but their own; the cause of cachin-
nation is the same, simple ignorance." *

Whether " snake-stones " be the true Bezoar or the
factitious animal charcoal, the principal of action is
much the same; both are absorbents, and both
chiefly consist of phosphate of lime. The first object appears to be inducing the blood to flow freely
to the wound, and then the remedy is applied. It
is very much like sucking out the poison; and, of
course, the sooner this is done after the wound is inflicted the better. After the virus becomes disseminated through the blood, it is useless to suck at the
portal by which it entered. In cases of poisoning
by the bite of a viper, cupping, and the application
of leeches have been effectual; such remedies, however, would be insufficient against the poison of the
more noxious tropical reptiles.

We are not aware that the " stones " alluded to are
worn as charms, amulets, or preservatives against
the bites of venomous serpents, but such things are
not uncommon in Eastern countries. As the teeth
of a tiger are sometimes worn as a charm against
attack from that animal, so perhaps the fangs of a
serpent may be regarded as a preservative against
the venom of serpents themselves. It is a current
belief amongst the natives in some countries where

* *The Zoologist*, p. 6983.

serpents abound, that anyone swallowing the contents of the poison apparatus of venomous snakes is
thereby preseserved from any ill effects accruing
from the bite of a serpent of that particular species.

The "cobra stone" is another romance which is
current in Ceylon and India. This curious story,
which is known throughout India, is to the effect
that some cobras, perhaps one in twenty, are in possession of a precious stone which shines in the dark.
This stone, according to the natives, the snake is in
the habit of carrying about in its mouth, regarding
it as a treasure, and defending it with its life. At
night the cobra deposits the stone in the grass and
watches it, as if fascinated, for hours, but woe to him
who then approaches, for the cobra is never more
dangerous than when occupied in this manner.
Finding that some of my Ceylon friends credited this
superstition, as I then regarded it (writes Professor
H. Hensoldt), I determined, if possible, to solve the
mystery. I offered five rupees to any coolie on the
estate who would bring me one of these cobra stones;
and one evening a Tamil came in hastily, to say
that he would show me the snake and its stone if I
would follow him. Without delay I went with him
to a little waterfall, distant over a mile from the
house. Close to the water's edge stood an immense
tamarind tree, and within fifty yards of it the coolie
halted, and mysteriously pointed to the root of the

tree. There the naja was to be found, but my guide
refused to go an inch further. As I could see nothing
from where I stood, I slowly and cautiously approached the tree, until, at about fifteen yards distance from it, I stood as if rooted to the spot. A
foot from the trunk I observed in the grass a
greenish light, apparently proceeding from a single
point. After a time I could see the cobra coiled
near the foot of the tree, slowly swaying its head to
and fro in front of the shining object. Save that
this shining light was steady and not intermittent, I
might at first have thought it due to the female of
the well-known fire-fly, for the air was swarming
with these insects. Unfortunately I had no gun, and
my guide, who seemed to feel that he was responsible
for my safety, entreated me so earnestly to let the
snake alone that I acceded. Moreover, he promised
to bring me this stone within three days, for he said
that the cobra, if not molested, would return to the
same spot night after night. The coolie kept his
word, for the second morning afterwards he brought
me the stone. He had climbed the tamarind tree
before dark, and after the snake had taken up its
position he had emptied a bag of ashes upon the
stone. The frightened reptile, after chasing about
for a while trying to find its treasure, had gone off.
The coolie remained in his safe position until day-
light, when he descended the tree, dug the stone
out of the ashes, and here it was in my hand. The

"cobra stone" was a semi-transparent, water-worn pebble of yellowish colour about the size of a large pea, which in the dark, when previously warmed, emitted a greenish, phosphorescent light. I found it to be chlorophane, a rare variety of fluor spar. The mystery is not difficult to explain. Cobras feed on insects, and seem to have an especial liking for fire-flies. I have often for hours watched the snakes in the grass catching the fire-flies, darting about here and there, a process which requires considerable exertion. Only the male fire-flies fly about, and a close observer will notice that a constant swarm of the male insects will fly near the females, which sit on the ground and emit an intermittent glowing light. The cobra uses his phosphorescent stone as a decoy for the fire-flies. No doubt the snake made the discovery by accident, night after night, perhaps, noticing how the fire-flies gathered about the shining pebble. Several snakes gathered, and it would require no great reasoning powers for the cobra to learn that the position of vantage was that nearest the pebble. Competition would lead to the snake's seizing and carrying off the treasure, and habit has become hereditary.

Another kind of " snake-stones," adder-gems, *ovum anguinum* or snake eggs, enter into the ancient superstitions of our own country. Borlase tells us *

* " Antiquities of Cornwall," p. 37.

C 2

that "in most parts of Wales, and throughout all
Scotland, and in Cornwall, we find it a common
opinion of the vulgar that about Midsummer-eve
(though in the time they do not all agree) it is usual
for snakes to meet in companies, and that by joining
heads together and hissing, a kind of bubble is
formed, which the rest, by continual hissing, blow on
till it passes quite through the body, and then it
immediately hardens, and resembles a glass ring,
which whoever finds shall prosper in all his under-
takings. The rings thus generated are called
Gleinau Nadroeth; in English, snake-stones." In
winter the viper may often be found in its hybernaculum, several individuals together, intertwined
and in almost torpid state. From this circumstance
probably some of the notions connected with the
stones alluded to may have been derived. Mason, in
his "Caractacus," puts into the mouth of a Druid the
following passage :—

> The potent adder-stone
> Gender'd 'fore th' autumnal moon ;
> When in undulating twine
> The foaming snakes prolific join ;
> When they hiss, and when they bear
> Their wondrous egg aloof in air ;
> Thence, before to earth it fall,
> The Druid, in his hallow'd pall,
> Receives the prize,
> And instant flies,
> Follow'd by th' envenom'd brood
> Till he cross the crystal flood.

The glass beads, in more recent times employed as charms, were used as a substitute for the rare "snake-stones." These "beads are not unfrequently found in barrows,* or occasionally with skeletons whose nation and age are not ascertained. Bishop Gibson engraved three: one of earth enamelled with blue, found near Dolgelly, in Merionethshire; a second of green glass, found at Aberfraw; and a third, found near Maes y Pandy, Merionthshire;"† Some have affirmed that in Cornwall, where they retain a respect for such amulets, they have a charm for the snake to make the "milprev," as it is termed, when they have found one asleep, and stuck a hazel wand in the centre of her spiral. "The country people," says Dr. Borlase, "have a persuasion that the snakes here breathing upon a hazel wand produce a stone ring of blue colour, in which there appears the yellow figure of a snake, and that beasts bit and envenomed, being given some water to drink wherein this stone has been infused, will perfectly recover of the poison."

We will leave Pliny alone‡ with his *ovum anguinum*, and the various other authors who have referred to these amulets, and proceed to matters of fact rather than of poetry and romance, concluding this chapter with a copy of the figures of "snake-

---

* Stukeley's "Abury," p. 44.
† Brande's "Popular Antiquities," iii., p. 371.
‡ Nat. Hist., lib. xxix., c. 12.

stones" given by Pennant in his " British Zoology,"
and who says of them—"Our modern Druidesses seem
not to have so exalted an opinion of their powers.
using them only to assist children in cutting their
teeth, or to cure the chincough, or to drive away
an ague."

ADDER-STONES.

THE COMMON LIZARD.

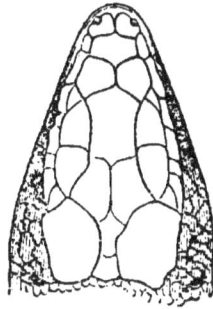

(*Zootoca vivipara.*)

Enlarged view of the upper surface of the head, showing the form, and arrangement of the plates.*

THE scaly or viviparous Lizard (Plate I.) is a common inhabitant of heaths and banks in elevated districts in England and Scotland, whilst it appears to have been one of the few reptiles which St. Patrick permitted to remain within the limits of the Emerald Isle. To what circumstance such forbearance in this instance is due, chroniclers seem to have been unable to tell. In some seasons, as for instance in

* Milne-Edwards in "Ann. des Sc. Nat.," ser. i., vol. xvi. t. 5, f. 5.

1860, this lizard is found in vast numbers everywhere in the county of Down. The observer who records its appearance in such plenty on the above occasion remarks, as a singular circumstance, that they never occurred there before, except a single individual at a time, and those at long intervals.[*] Lord Clermont observes that is is never found in low countries, but frequents mountain districts in the greater part of North and Central Europe, and is common in Switzerland, Germany, Poland, and France, as well as in Scotland, England, and Ireland. In Italy it is only found in the Alpine regions of the North, and it also inhabits the hilly parts of Belguim and Russia.[†]

This lizard differs in a most important point from the other species to be mentioned, and on this account has been placed by naturalists in a new genus called *Zootoca*, from the Greek word *zoos*, "life," and *tokos*, "offspring," on account of its young bursting through the very thin membrane-like covering of the egg at the time of birth, and are, therefore, ovo-viviparous; in which feature this lizard resembles the viper. The young, as soon as they are born, have the free use of their limbs, and run about in company with their parent, soon com-

---

* *The Zoologist*, p. 7172.
† Clermont's "Quad. and Rept. Eur.," p. 184.

mencing the collection of insects on their own
account, impelled by the feelings of hunger. The
specific name of *vivipara* has a Latin origin, with a
like meaning to the generic. This lively little reptile will often be found sunning itself between spring
and autumn. All its movements are exceedingly
graceful and vivacious. In an instant it darts upon
an insect coming within its range, which as speedily
disappears down its throat. Like the slow-worm
and sand lizard, it is often the object of persecution,
though itself perfectly harmless; but, on account of
its reptile form, does not escape calumny. During
summer the pregnant female may be discovered
basking in the direct rays of the sun, and is then far
less willing to be disturbed than at other seasons.
Tradesmen who supply materials for acquaria, fern-
cases, and domestic vivaria, tell us that while they
find a ready sale for the newts, there is such small de-
mand for lizards, because "people are afraid of them;"
that they seldom keep any on hand, but collect and
supply them to order. This is a foolish prejudice,
because they would make an aggreeable addition to
the attractions of a fernery, and assist in keeping it
free from insects. Their movements are more
graceful and rapid than those of the newts, and
certainly would not require any larger amount of
attention.

This species is smaller than the sand lizard, not

exceeding from five and a half to six inches in length. The tail is longer in proportion, and of a different shape, retaining the same thickness for the first half of its length, and then diminishing gradually to its extremity; the palate is without teeth, the temple is covered with small polygonal plates, with a large angular one in the centre. The scales on the back are long, narrow, and hexagonal, and less distinctly keeled than in the next species. The head is more depressed and the nose sharper. The plates of the belly are in six rows, with two small marginal series ; the preanal plate is bordered by two rows of scales. The fore legs reach to the eye, the hind legs extend along two-thirds of the sides ; pores from nine to twelve on each thigh. The back is brown, olive, or reddish, with a black band on each side from the head to the tail; a second dark band runs along the side, and is edged with white. The under parts are spotted with black upon a whitish ground, generally with a bluish or greenish tinge.*

The relative size and vivaparous character are the best features whereby to distinguish this species from the next.

* Lord Clermont's " Reptiles of Europe," p. 184.

THE SAND LIZARD.

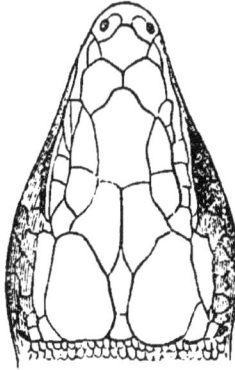

(Lacerta agilis.)

Enlarged view of the upper surface of the head, showing the form and arrangement of the plates.*

THIS and the latter species were long confounded together. It appears to vary considerably, both in colour and size, is generally larger, but not so common as the Scaly Lizard (Plate 1). It seems to be pretty widely distributed over Europe, being found almost everywhere except in the extreme North; confined more especially to lowland districts. It is said to

* Milne-Edwards in "Ann. des Sc. Nat.," ser. i. vol. xvi. t. 5. f. 4.

abound in Germany, Switzerland, Poland, Northern
Russia, in Siberia, and generally through Central
Europe. It appears to occur freely in the neighbourhood of Poole, but we do not remember to
have met with records of its occurence in the North
of England, or in Scotland, although it may have
been confounded with the foregoing, especially as it
is evidently a Northern rather than a Southern
species.

Unlike the Scaly Lizard, which is in reality the
common lizard with us, it is oviparous, the female
laying twelve or more eggs in the sand, and leaving
them to be hatched by the heat of the sun. She
hollows out a cavity, or rude nest, for the purpose,
and covers her eggs with the sand. It possesses a
"snappish" temper, is not readily domesticated, and
refuses food under confinement. Like other reptiles,
it passes its winters in a state of repose.

The liver, bile, excrements, and eggs of the
Lizard were in former times employed as remedies
in certain diseases, and the entire animal has been
proposed as a substitute for the Scink, a reptile
allied to the Lizards, which had a great reputation
in olden times, the use of which has very recently
been attempted to be revived. Dr. Gosse, of
Geneva, has maintained that the ancients were
justified in employing the Scink in medicine, inasmuch as it possesses powerful stimulant and sudorific

properties, which might be usefully employed in various diseases.*

In Mr. P. L. Simmonds's interesting and amusing "Curiosities of Food," several species of reptiles are enumerated as affording food to the natives in the various countries in which they are found. Of these, the Iguana holds one of the chief places in public esteem. This is a gigantic lizard found in many tropical countries, where it attains a length of three feet, and has flesh "which is reckoned as delicate as chicken, and but little inferior to turtle in flavour." Humboldt remarks that in inter-tropical South America, all lizards which inhabit dry regions are esteemed delicacies for the table. The gigantic crocodile, alligator, gavial, and cayman, are also served at repasts. This, however, is somewhat foreign to our subject, and all who are interested in reptilian delicacies we must refer to the book in question.

The fossil Saurians of bygone ages were the giants of those days. Dr. Mantell thought it probable that the largest iguanodons may have attained a length of from sixty to seventy feet. The Labyrinthodons were also of considerable size. Remains and traces of numerous reptiles have been found in the strata of our own islands. "Impressions of the feet

* See Moquin Tandon's "Medical Zoology," p. 69.

of a Labyrinthodon were found by the late Mr. Hugh Strickland in the lower Keuper sandstone of Shrewly Common, Worcestershire; and bones and teeth have been discovered near Kenilworth, in the Permian sandstones of the geological surveyors, as well as in the upper Keuper beds. Five species of Labyrinthodont reptiles have been found in Great Britain, all of which must have been very unpleasant-looking animals, with fearful jaws, adapted especially for biting. And yet such animals lived on the shores of a sea-bed which now constitutes much of the pleasant vales of Worcestershire and Cheshire, the shores of the New Red Sandstone Sea.*

Returning from "stewed Iguana" and extinct reptiles to the little Lizard of the present degenerate days, we may observe, that though neither formidable in size, repulsive in appearance, nor in any sense aggressive or noxious, it has many enemies, some amongst bipeds with feathers, and some amongt bipeds without. It has personal interest in the "smooth snake," and its proximity to its own locality; for that reptile has a great predilection for a lizard at luncheon, whilst the common snake prefers a frog or a newt. But the most relentless persecution is carried on by schoolboys, and even adults, with more zeal than discretion,

---

* "Old Bones," by the Rev. W. S. Symonds, p. 81.

who fancy that they are doing "the state some service," in efforts to accomplish its extermination. That it is not only inoffensive, but useful in keeping up the balance of nature, by its reduction of insect life, its legitimate prey, is a truth, like the "small bird question," which may be acknowledged when it is all but too late.

There are two varieties of this lizard, indicated by their relative colour. In one the general tint is brown; in the other it is green. The most common variety has the back of a sandy-brown colour, sometimes spotted with black, with the sides greenish in the male, but brownish in the female; the belly is white and often spotted. In length it is from seven to nine inches, of which the tail occupies more than half, or nearly two-thirds. The scales of the upper part of the body are roundish or angular, and distinctly keeled; the plates of the belly are arranged in six rows, of which the two central rows are the narrowest. The tail is covered with from fifty to eighty distinct whorls or rings of scales, which are longer and narrower than those of the back. It is thicker and rather more clumsy than the last species, and the limbs are stouter and stronger, and it is less graceful and vivacious in its movements. There are other and more minute points of difference, but these are of interest rather to the zoologist than to the general reader. It may

be observed, however, that in this species there are
to be found, in addition to the ordinary teeth at the
margin of the upper and the lower jaws, also a few
very small ones seated on the back part of the
palate, and which are wanting in the common lizard.

Professor Bell states on the faith of a gentleman
of his acquaintance, that the brown varieties are
confined to sandy heaths, the colours of which are
closely imitated by the surface of the body, and
that the green variety frequents the more verdant
localities. This, he adds, he had not been in a
position either to refute or confirm, and could only
vouch for the existence of two such varieties, at a
comparatively short distance from each other.

PLATE II.

1. Green Lizard.   2. Blindworm.

LITH. IN HOLLAND BY ERK. K & BINGER, 21ST LEFNERS ST LONDON

THE GREEN LIZARD.

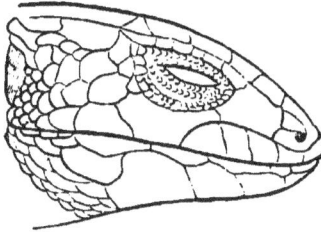

(*Lacerta viridis.*)

Enlarged view of the side of the head, showing the form and arrangement of the plates.*

Is the Green Lizard (Plate 2) really a native of Britain? That is the very knotty question which we desire to settle, both for our own satisfaction and that of our readers, but cannot divest our minds of a lingering doubt whether it may not have descended to us in a similar manner to the shower of edible frogs which Mr. Penney rained down upon Foulmire. That this species has been found in Great Britain, in an apparently wild state, is without a doubt, but how it came there is past finding out.

* Milne-Edwards in "Ann. des. Sc. Nat.," ser. i., vol. xvi., t. 7, f. 2.

D

If we turn over the pages of the earlier volumes
of the Zoologist we here and there encounter little
facts of a very stubborn nature, relative to the Green
Lizard, in every instance guaranteed by some name
well known in the annals of science. One of the
earliest of these notes is by Dr. Bromfield,* in which
he states, " I am told, on competent authority, that
Lacerta viridis is quite frequent, and even abundant, in the neighbourhood of Herne Bay. I may
add, there can be no doubt about the species, and
that it certainly is not the smaller green lizard of
Poole, but identical with the species long known to
inhabit Guernsey, as my friend, Professor Bell, has
received a specimen from Herne Bay, but not in
time to notice the discovery for the second edition
of his 'British Reptiles.'" Corroborative of this,
and on the same page of that journal, occurs the
following communication from the late Mr. John
Wolley :—" Seven or eight years ago, a schoolfellow
of mine at Eton, a native of Guernsey, assured me
he had seen lizards in Devonshire precisely similar
to the lizards of his own island;" and again, "Nearly
two years since a learned Professor of the University
of Edinburgh mentioned that he had dissected a
Green Lizard brought by a botanical party from the
Clova Mountains." Still later another credible wit-

* Zoologist, p. 2707.

ness, the Rev. W. H. Cordeaux, gives evidence to
the effect that he had examined the Reptilia of the
Canterbury Museum, and had there found a male
and female of this species, but that no trustworthy
information could be obtained of the locality and
date of their capture.* In 1863 a paper was read
and a specimen exhibited before the Holmesdale
Natural History Club, at Reigate, by Mr. J. A.
Brewer. The specimen was caught by a labourer on
a bank by the side of the road, a little way from
Dorking, on the road to Reigate, and purchased of
him the same evening. In the course of this paper,
Mr. Brewer remarks :†—"The occurrence of this soli-
tary specimen is not sufficient in itself to establish
it as a British species ; but on showing it, a few days
since, to Mr. John E. Daniel, a well-known naturalist,
and who certainly would not have been likely to
mistake this beautiful species, he informed me that
a few years since he had observed three or four
specimens of it on the heath, about half-a-mile south
of Wareham, in Dorsetshire, one of which he captured, and is quite certain of its identity with the
Green Lizard (*Lacerta viridis*), a species which
he is well acquainted with, having frequently seen
it in Germany, and received specimens from the
Channel Islands. Gilbert White, in his "Natural

* *Zoologist*, p. 2855.   † Ibid, p. 8639.

D 2

History of Selborne," has the following remark, which probably applies to this species :—" I remember well to have seen formerly several beautiful green *Lacerti* on the sunny sand-banks near Farnham, in Surrey, and Ray admits there are such in Ireland." All we can say to such evidence is, that the facts are too strong and well attested to deny that the Green Lizard has been several times found in this country; and though it is quite possible, nay, *certain*, that many a time and oft they have been brought from Guernsey, and turned adrift here, they are at least naturalized, and deserve a notice in any history of "Our Reptiles." Some of the more recent captures may have been individuals from the " imports," but it is equally probable that the Green Lizards mentioned by White were really indigenous, and the same species. Who shall determine satisfactorily that they were not?

The large Lizard quoted by Pennant can scarcely be the same as the present, unless the length was much exaggerated. He thus records it :—

The most uncommon species we ever met with any account of, is that which was killed near Woscot, in the parish of Swinfoid, Worcestershire, in 1741, which was two feet six inches long and four inches in girth. The fore legs were placed eight inches from the head, the hind legs five inches beyond those; the legs two inches long, the feet divided into four toes, each furnished with a sharp claw. Another was killed at Penbury, in the same county. Whether these are not of exotic descent, and whether the breed continues, we are at present uninformed.

This reptile is found inhabiting the greater part
of Central and Southern Europe, including France
as far north as Paris. It is very common in the
south of that country, all over Italy and the south
of Switzerland; is found in Sicily, Greece, Poland,
Austria, the Crimea, and Barbary.*

This is a really fine lizard, large, beautiful, and
attractive, at least as much so as a reptile can be.
Its entire length seldom exceeds fifteen inches,
though sometimes attaining eighteen inches in the
Morea. Although the colouring is very variable,
green is a prevailing tint. The upper surface is
sometimes of a uniform green, at others green
with yellow spots, and occasionally brown with green
or white markings, rarely brown with white streaks
edged with black. The lower surface is usually
yellow. It occurs on hedge-banks and grassy
places.

It is of a rather tractable nature, submits to confinement, and ultimately becomes familiar, for which
reason it is not an uncommon pet in the warmer
countries of Europe, where it is chiefly found; and
is occasionally imported into this country for a like
purpose. If it really cannot be proved to be truly
indigenous to Great Britain, all we can say is, " the

* Lord Clermont's " Quadrupeds and Reptiles of Europe,"
p. 185.

more's the pity," because it would be a handsome addition to the reptile fauna of any country, and being inoffensive and tractable, could be welcomed without regret.

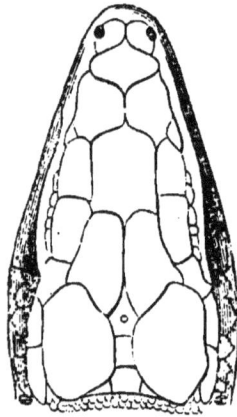

Enlarged view of the upper surface of the head of the Green Lizard (*Lacerta viridis*) showing the form and arrangement of the plates.—Milne-Edwards, t. 5, f. 3.

THE BLINDWORM.

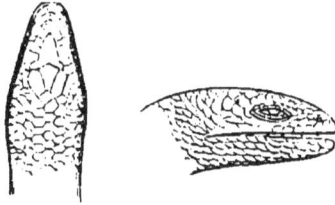

(*Anguis fragilis.*)

In external form, the Slow-worm so much resembles a little snake that those of our readers who have not delved deeply in the mines of Reptilian knowledge will, at first, evince surprise at finding it amongst the Lizards, in company with whom it is placed by scientific men. But as much as first appearances are opposed to this union, a closer examination will prove that it is "the right thing in the right place." The most convincing proof of this is to be found in the fact that, although not possessed of external legs, the rudiments of these organs are concealed beneath the skin. The eyes, again, are furnished with moveable eyelids—a phenomenon not occurring amongst the snakes, but found in the

lizards. Then, again, the jaws of serpents are so
constructed as to expand sufficiently to admit as
large a body as will pass down into the stomach,
whilst in the Blindworm this is not the case. The
tongue, also, is notched at the point, but not cleft or
forked, as in the Ophidians. Finally, the back, belly,
sides, and tail are all covered with small rounded
scales, which, as we shall see shortly, is by no
means the case with serpents. For these reasons,
and some others too technical to deserve a place
here, the Blindworm, or Slow-worm, is classed
with lizards, and is, in fact, a lizard without visible
legs.

This anomalous reptile is found all over Europe,
except the most northern countries, and, again, we
must except Ireland; but in England and Scotland
it is very common. Being less susceptible of cold,
it is found further north, and comes out from its
hybernaculum earlier than most other reptiles. Like
the snakes, it casts its slough, which it leaves behind, and does not attempt to devour. This is
generally turned "inside out," as in the snake and
viper; but the tail portion is sometimes excepted;
out of this the tail is sometimes drawn without the
skin being reversed. When in confinement the
slough usually comes off in fragments.

Like the viper and the scaly lizard, the Blindworm is ovo-viviparous—*i.e.*, its young are brought

forth alive, from seven or eight to ten or twelve
at a birth.

There is one peculiarity in this reptile, which is
met with in no other British species; whence its
specific name of *fragilis* is derived. When attacked,
alarmed, or taken hold of, it becomes rigid; and in
this condition any effort to bend it results in breaking off a portion of the tail; a slight blow will sever
it in the same manner; and when taken hold of by
the tail it will sometimes make its escape, leaving
that extremity in the hand. Within certain limits
it has also the power of replacing the broken part;
for a short conical tip at length occupies the place
of the severed tail. It is very inoffensive, quiet,
timid, and retiring in its habits; and is, in fact, a
model of reptile virtues, without possessing any taint
of reptile vices. Like some friends, it "improves
upon a closer acquaintance."

Mr. George Daniel has given us some of the most
complete observations on the habits of this reptile :—
"A Blindworm that I kept alive for nine weeks
would, when touched, turn and bite, although not
very sharply; its bite was not sufficient to draw
blood, but it always retained its hold until released.
It drank sparingly of milk, raising the head when
drinking. It fed upon the little white slug, so
common in fields and gardens, eating six or seven of
them, one after the other; but it did not eat every

day. It invariably took them in one position. Elevating its head slowly above its victim, it would suddenly seize the slug by the middle, in the same way that a ferret or dog will generally take a rat by the loins; it would then hold it thus sometimes for more than a minute, when it would pass its prey through its jaws, and swallow the slug, head foremost. It refused the larger slugs, and would not touch either young frogs or mice. Snakes kept in the same cage took both frogs and mice. The Blindworm avoided the water ; the snakes, on the contrary, coiled themselves in the pan containing the water which was put into the cage, and appeared to delight in it. The Blindworm was a remarkably fine one, measuring fifteen inches in length. It cast its slough whilst in my keeping. The skin came off in separate pieces, the largest of which was two inches in length, splitting first beneath, and the peeling from the head being completed the last."*

As will be seen on comparing this extract with some of our own observations, it does not entirely accord in all its minor details. For instance, in its food ; for, although it is not at first disposed to eat in confinement, it becomes at length more sociable, and will then eat earthworms as freely as slugs.

* Note to White's " Selborne," Bennett's edition.

Of all the guiltless beings which are met with, we
have none less chargeable with criminality than the
poor Slow-worm; yet none are more frequently destroyed than it, included, as it is, in the general and
deep-rooted prejudice attached to the serpent race.
The viper and snake, though they experience no
mercy, escape often by the activity of action; but
this creature, from the slowness of its movements,
falls a more ready victim. We call it a " blindworm,"
possibly from the supposition that, as it makes little
effort to escape, it sees badly; but its eyes, though
rather small, are clear and lively, with no apparent
defect of vision. The natural habits of the slow-
worm are obscure; and living in the deepest foliage
and the roughest banks, he is generally secreted
from observation ; but loving warmth, like all his
race, he creeps, half torpid, from his hole, to bask in
spring-time in the rays of the sun, and is, if seen,
inevitably destroyed. Exquisitely formed as all
these gliding creatures are, for rapid and uninter-
rupted transit through herbage and such impedi-
ments, it is yet impossible to examine a Slow-worm
without admiration at the peculiar neatness and
fineness of the scales with which it is covered. All
separate as they are, they lap over and close upon
each other with such exquisite exactitude, as to appear only as faint markings upon the skin, requiring
a magnifier to ascertain their separations; and to

give him additional facility of proceeding through rough places, these are all highly polished, appearing lustrous in the sun, the animal looking like a thick piece of tarnished copper wire. When surprised in his transit from the hedge, contrary to the custom of snake or viper, which writhe themselves away into the grass in the ditch, he stops, as if fearful of proceeding, or to escape observation by remaining motionless; but if touched, he makes some effort to escape. This habit of the poor slow-worm becomes frequently the cause of his destruction.*

This species is generally about ten inches in length, and rarely exceeds fourteen inches. Its general colour on the upper surface is of a brownish grey, with a silvery or bright steel-like appearance. There are commonly several parallel rows of minute darker spots along the sides, and one down the middle of the back. Underneath it is of a bluish black, with a whitish network. The young at first are whitish, then of a light yellowish grey above, with a black line running down the back, and with black bellies. The mature reptile is cylindrical, or slightly squared, in form, gradually decreasing towards the tail, which ends abruptly. The latter often equals the body in length, and is covered

* Knapp's "Journal of a Naturalist," p. 309.

above and below, as well as the body, with small, rounded, closely-fitting scales. The eyes are small, and provided with moveable eye-lids. The tongue is broad, but not very long, and notched at the tip; and the teeth are minute, and slightly hooked.

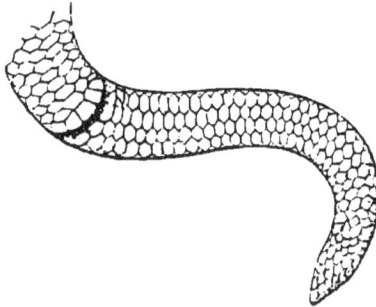

THE TAIL.

THE COMMON SNAKE.

(*Tropidonotus natrix.* Dum. & Bibr.)

A mass of Snake's eggs from a dunghill.

EVERYBODY involuntarily shudders at the name of a snake. Very few possess courage enough to attempt staring one out of countenance, or staying to count the number of scales on its head. Fancy oneself deeply intent, with nose unusually low, seeking the ruddy wild strawberry on a sunny· hedge-bank, and even whilst smacking the lips with the relish of the

PLATE III

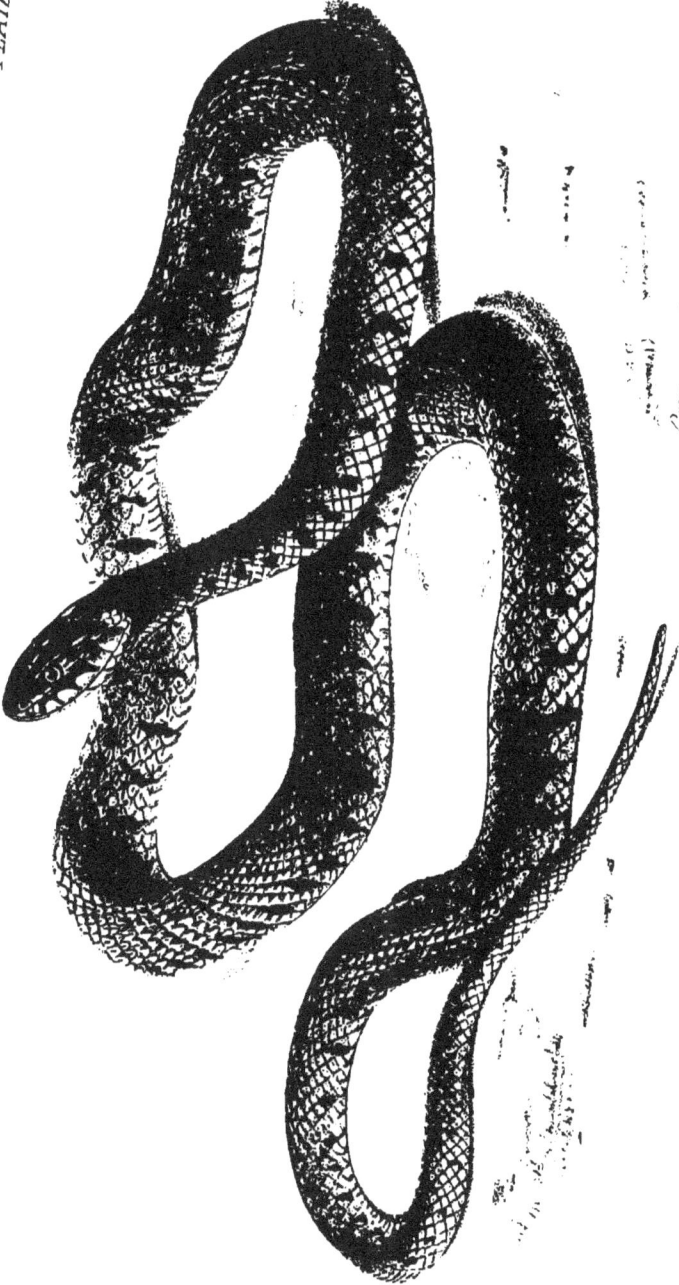

Common Snake.

LIFE IN HOLLAND BY EMRIK & BINDER, 21 BERNERS ST LONDON

tart little fruit but lately conveyed there, about to
pluck another yet larger and redder, when lo! beneath our very fingers glides the sleek, attenuated
form of the reptile—ay, within ten inches of our
depressed nose. Under such circumstances, should
we be surprised at finding ourselves starting back;
at feeling a slight and momentary sensation, as of a
drop of water trickling down our back ; or at for-
getting to observe whether the intruder was really a
viper or a snake?

What is called the Common (Plate 3.) or Ringed
Snake with us, is really the most common snake in
Europe. It is found in almost every country from
Sweden in the North, to Sicily in the South, excepting Ireland, whence it is said to have been
banished by the good Saint Patrick, and in the
extreme north of Russia, whence it is probably
excluded by the temperature, independent of saintly
proscription. In Britain it is far more common
in the South than in the North; some have
even had the temerity to deny its occurence in
Scotland, but apparently without good foundation.
Attempts have been made to introduce it into
Ireland, an account of one such being narrated in
Bell's "British Reptiles." But the hand of every
true Irishman was armed against the innovation, and
the imports soon suffered extermination.

However, one may shudder at the sight of a snake,

this species is perfectly harmless, indeed rather tractable under confinement, and certainly, in common with the rest of its tribe, exceedingly graceful in its undulations, and possessed of a truly fascinating eye. As we write this paragraph, a lively individual about two feet in length is gazing intently at the movements of our fingers, as if to divine therefrom whether any malignant libel is being penned, or whether the movements are those of flattery.

This species is truly oviparous. Its eggs, from sixteen to twenty in number, attached together by a glutinous secretion, are deposited in some favourable locality, as a dunghill, and are hatched by the heat developed, or that derived from direct exposure to the sun. In this circumstance it will be seen to differ from the viper. Knapp gives a good account of a nest of snake's eggs :—

My labourer, this July the 18th, in turning over some manure, laid open a mass of snake's eggs, fifteen only; and they must have been recently deposited, the manure having very lately been placed where they were found. They were larger than the eggs of a sparrow, obtuse at each end, of a very pale yellow colour, feeling tough and soft, like little bags of some gelatinous substance. The inner part consisted of a glareous matter like that of the hen, enveloping the young snake, imperfect, yet the eyes and form sufficiently defined. Snakes must protude their eggs singly, but probably all at one time, as they preserve no regular disposition of them, but place them in a promiscuous heap. At the time of protrusion they appear to be surrounded with a clammy substance, which, drying in the air, leaves the mass of eggs united wherever they touch each other. I have heard of forty eggs being found in these deposits ; yet, notwithstanding such provision for multitudes, the snake, generally speaking, is not a very common animal.*

Having deposited her eggs, the female appears to have no further care or thought of her progeny. She neither watches over them to preserve them from injury, nor, when hatched, does she take them under tuition in the torturous policy of a snake's existence. The incubation of snakes was a knotty point widely discussed when the python of the Zoological Gardens laid eggs, which were never hatched.

This snake, in common with others, changes its skin at intervals, but not, as has been stated, at regular periods, or once a year ; but sometimes four or five times during the year, and often less, according to circumstances. In this "sloughing" process the reptile bursts the cuticle about its neck, draws out its head, the old skin is thrust back, and the snake crawls out. In this process the skin is turned inside out, and left on the grass to scare unwary females into the belief that they have seen a snake, little dreaming that they have only been shuddering at its old clothes.

What does the snake eat?—Undoubtedly it delights in frogs, young birds, birds' eggs, and even mice. Imagine the little shudder and start in which

* Knapp's "Journal of a Naturalist." p. 306.

E

we indulged in boyhood, on putting our hand, with
felonious intent, into a bird's nest (we couldn't see
into it), and finding our fingers come in contact
with the smooth cold folds of a coiled-up snake! It
was the last time we *felt* for eggs before *seeing* them.
That was an experiment too satisfactory in its results to require repetition.   The author already
quoted gives an interesting account of a snake's
meal :—" If it be a frog, it generally seizes it by the
hinder leg, because it is usually taken in pursuit.
As soon as this takes place, the frog ceases to make
any struggle or attempt to escape.   The whole body
and legs are stretched out, as it were, convulsively,
and the snake gradually draws in, first the leg he has
seized, and afterwards the rest of the animal, portion
after portion, by means of the peculiar mechanism
of the jaws, so admirably adapted for this purpose.
When a frog is in the process of being swallowed in
this manner, as soon as the snake's jaws have reached
the body, the other hind leg becomes turned forwards; and as the body gradually disappears, the
three legs and the head are seen standing forwards
out of the snake's mouth, in a very singular manner.
Should the snake, however, have taken the frog by
the middle of the body, it invariably turns it, until
the head is directed towards the throat of the snake,
and it is then swallowed, head foremost."   The frog
is not only alive during the above process, but often

after it has reached the stomach. Mr. Bell says, " I once saw a very small one, which had been swallowed by a large snake in my possession, leap again out of the mouth of the latter, which happened to gape, as they frequently do immediately after taking food."

During the present summer a gentleman of our acquaintance saw a lad kill a snake in a wood. It was a very large one, and the boy cut it open along the under surface with his pocket-knife. By this means a full-sized frog was liberated from the stomach of the snake. It was very lively, and soon hopped away. Why may not young vipers remain as long with equal ease in the stomach of their parent ?

The snake is very fond of the water, and may often be surprised, coiled up in sunny weather, with its head out enjoying the luxury of a bath. It will dive after the water-newts, especially when rather hungry, bringing them to the shore in its mouth, and devouring them upon dry land. Some kinds of snake have been detected catching fish ; but whether this was merely an idiosyncrasy on the part of one or two individuals, or whether it is a confirmed habit, we are not in possession of sufficient evidence to determine. Such a predilection on the part of the common snake we have not yet heard of. This reptile is generally found in wet situations, or not far from

water, whilst the viper evidently prefers a drier
locality. We have heard of vipers being found on
marshes, but are doubtful whether the creatures so
called were not common snakes.

The entire length of this species is generally over
two feet, sometimes exceeding three feet, and rare
instances have been recorded, of which several occur
in the pages of the *Zoologist*, of its reaching, and
even exceeding, four feet. The female is usually
the largest. The colour of reptiles is very prone to
variation, but the tint of the upper surface in this
species is generally of a brownish-grey with a greenish tinge, sometimes nearly that of an ash stick.
Along the back are two parallel rows of small black
spots, with a series of larger blotches of black of
variable sizes along each side. The under surface is
of a dull lead-colour, sometimes mottled. The scales
and their arrangement, especially on the head, differ
from those of the viper. In the common snake the
head is covered by large plates, of which there are
three between the eyes, and those above and below
are arranged in pairs. The scales along the back
are oval and distinctly keeled, with those of the
sides broader, and less keeled. On the belly the
plates are single, extending from side to side, and
about one hundred and seventy in number; but
in this there is also variation. The under plates of
the tail are in pairs. Behind the head is a broad

collar or a pair of spots of a bright yellow colour, behind which are black spots. The teeth are in two rows in both upper and lower jaws, and the tongue is forked to one-third of its length.

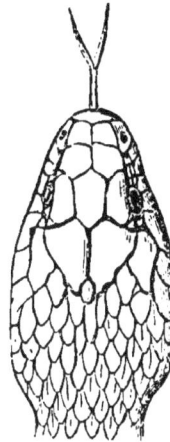

VIPER.   SNAKE.

THE SMOOTH SNAKE.

(*Coronella lævis.* Boie.)

THERE can be little doubt that this reptile has been
occasionally found in this country for many years,
although it is only recently that it has received the
name to which it appears to be entitled, and there-
with its true place in our reptile fauna. It is somewhere about half a century since Mr. Simmons
caught a curious little snake near Dumfries, which
appeared to differ so much from the common ringed
snake that Sowerby figured it in his "British Mis-
cellany,"* and called it the Dumfries snake (*Coluber
Dumfrisiensis*). After this, the account and a copy
of the figure appeared in Loudon's "Magazine of
Natural History."† When Professor Bell published
his account of the British Reptiles, he gave these
particulars, but at the same time expressed a doubt

* Sowerby's "Miscellanies," iii., t. 3. † Vol. II., p. 428.

PLATE IV

Smooth Snake.

whether the specimen was not an immature one of
the common species.* As the specimen is not in
existence, it cannot be positively affirmed what it
was; but, from the description, there is reason to
believe it was really the first recorded capture of the
snake whose name stands at the head of this chapter. " It was about three or four inches in length,
of a pale-brown colour, with pairs of reddish-brown
stripes from side to side, over the back, somewhat
zig-zag, with intervening spots on the sides. The
abdominal plates were 162, those under the tail
about 80. The most remarkable peculiarity
mentioned, however, is, that the scales are extremely
simple, not carinated."

Towards the close of 1859 the Hon. Arthur
Russell sent to the British Museum a female specimen of the Smooth Snake (plate 4), which was taken
by a resident near the flagstaff at Bournemouth,
Hampshire, and Dr. Gray communicated a notice of
the fact to the *Zoologist* (p. 6731). Supplementary
to this notice, the editor added a description of the
reptile from Lord Clermont's work. In the next
number a communication appeared from Mr.
Frederick Bond, in which he stated :—" I captured
a specimen of the new British snake five or six years
ago, in June, near Ringwood, Hants. I thought at

* Bell's " Reptiles," 2nd ed., p. 60.

the time 1 had something new, but, not taking
much interest in the reptiles, it was put into spirits
and forgotten until I saw Dr. Gray's notice.   I have
sent this specimen to the British Museum, so that
anyone may see it." *

Nothing more was heard of the Smooth Snake in
Britain till the year 1862, when, between October
and December, several communications appeared in
the *Field*, and Mr. Buckland seemed to claim its
discovery as an addition to the British Fauna.   Mr.
Bartlett, of the Zoological Gardens, Regent's Park,
ultimately published an account, in which he
stated,†—" It was on the morning of the 24th
August, 1862, I saw for the first time one of these
animals, Mr. Fenton having stopped me as I was
driving along the road in the Regent's Park, and
taking from his pocket what I then thought was a
viper, asked me if I would accept it for the Zoological Gardens."   From the Proceedings of the
Zoological Society we learn that, on one occasion,
Mr. F. Buckland exhibited this specimen, and
ultimately‡ several others, which had been found
in this country.   So deficient, however, in all the
necessary details of date, place, and circumstances
of capture were these recent accounts, that, had
there not been prior and more satisfactory records

---

* *Zoologist*, p. 6787.   † *Intellectual Observer*, iii., p. 149.
‡ November 11th, 1862.

in existence, it would have been doubtful whether, on such incomplete details, it had been prudent to recognize the Smooth Snake as a British native. All who have devoted themselves to the study of the natural history sciences know how very essential it is that the evidence of the occurrence of any new, or supposed new species, should be complete and authentic. Deficiencies in these respects have heretofore caused infinite trouble, and too great prudence cannot be exercised in the endeavour to prevent their occurrence again. In this case we think the evidence sufficient, as above detailed, for the recognition of the present species as a true native.

The Smooth Snake inhabits central and southern Europe, is found in various parts of France, but is not very common in the south of that country: it occurs in Sicily and in the whole of Italy and its islands, but is more frequent in the north than in the south of that peninsula; it is included in the fauna of Galicia and the Bukovina, Silesia, and Carniola; it is common in Switzerland, near Zurich, but rare in Belgium, where it has been met with near Louvain, and on the right bank of the Moselle. Schinz states that it has been found in Sweden, but it is everywhere less abundant than the common snake.*

* Lord Clermont's "Quadrupeds and Reptiles of Europe."

In reference to the occurrence of this snake in
Sweden, Mr. Wheelwright, in a communication to
the *Field* newspaper,* says, " It is common on most
parts of the Continent, and by no means rare in
Sweden.   It is met with as far north as Upsala, but
nowhere more common than around Gothenburg.
We call it in Sweden the ' Slat Snok,' or Smooth
Snake ; and in this lies the principal distinguishing
mark between it and the common ringed snake.   In
the common snake the scales on the upper parts of
the body are imbricated, those on the back being
lancet-shaped, and distinctly keeled along the
middle, whereas in the Smooth Snake the scales are
oval, altogether smooth, without the slightest indi-
cation of a keel." As far as I can observe, with us
the Smooth Snake appears to be partial to stony
tracts ; it is perfectly harmless, and its principal
food appears to be mice ; and one which was kept a
long time in confinement would not touch a frog.
With us it rarely exceeds about two feet in length ;
it appears to be of a much tamer and more com-
panionable nature than the common snake.   A most
singular thing is that, according to Schlegel, the
female brings forth living young.   He says, " The
eggs take three or four months to hatch inside the
mother, and in the end of August she brings forth

* The *Field*, October 18th, 1862.

about twelve young, which are at first altogether
white." This latter point was definitely set at
rest by the fact of a snake of this species, caught in
the New Forest, giving birth to six young ones
whilst in confinement. Mr. Buckland gave a full
account, at the time, of the circumstances connected
with the history of this reptile progeny. " The old
mother snake is coiled up in a graceful combination
of circles, her little family are nestled together on
her back; they have twisted their tiny bodies
together into a shape somewhat resembling a double
figure of 8, and there they lie basking at their ease
in the mid-day sun. The old mother is vibrating
her forked tongue at me ; the little ones are imitat-
ing their mother's action, and are vibrating their
tiny tongues also ; the mamma's head is most
beautifully iridescent in the sun, and her babies are
in this respect nearly as pretty as their mother.
They are about five inches long, about as thick as a
small goose-quill, and smoother than the finest
velvet. Their eyes are like their mother's, their
tails are unlike their mother's; she has lost the tip
of her tail—her young ones have not, they are
tapered off to a point as sharp as a pin. Their
skins are of a brownish-black colour, and marked
like their mother's, only that these markings are
not yet well developed. The scales on the under
parts of their bodies are of a beautiful pale glitter-

ing blue; altogether they are real little beauties."*
Evidently the writer of these lines looked upon his
pets with the eyes of a connoisseur, and had they
been anything else but snakes no doubt the ladies
would have made a rush at him to have secured
pets to ornament their boudoirs. Amongst the
interesting communications which appeared at the
time relative to the Smooth Snake, was one from
Dr. Günther, our best authority in serpent lore.
" A large male specimen of this snake," he says,
" which I kept for a long time on account of its
tameness, fed exclusively on lizards, never on mice
or frogs. After having fed it for some time with
ordinary-sized lizards, proportionate to the size of
the snake, I brought a very large specimen of
Lacerta agilis to its cage, in order to try the
strength of the snake. The lizard was immediately
seized; but after a long fight, during which the
lizard several times appeared to be entangled in the
writhings of the snake, always managing, however,
to free its head which had been seized by the snake,
the latter changed the point of attack, and got hold
of the tail of the lizard. This, of course, broke off,
and was devoured by the snake. From this time
the snake always seized the tails of the lizards given
him for food, without further attacking them; nor,

* The *Field*, October, 1862.

if tailless lizards were put to him, would he attempt to devour them."*

Still more recently Dr. Opel has contributed to our knowledge of the habits of this snake, especially when under confinement.† From his observations it would appear that the ground colour for the six or seven days succeeding " sloughing " is of a beautiful steel-blue, which from that time gradually fades, until at last it settles into a dirty brownish-yellow. One of the specimens on which Dr. Opel's observations were made was captured by him in the Fürstensteiner Grund, near Salzbrunn, in Silesia, and carried to Dresden closely packed in a tin case. After a journey of eight days in this manner, it was in good health, and gave no signs of exhaustion. Minute particulars are given of three separate " sloughings," which took place during one year in the months of June, July, and August ; but he adds there can be no doubt that, in the wild state, the first sloughing commences in April. " Immediately upon losing its skin follows a desire for food ; at the same time the prisoner makes numerous attempts to escape, and is altogether restless and excited, though at other times it lies quietly in a corner of its cage." After alluding to the fact that it is of

---

* The *Field*, September 13th, 1862
† The *Zoologist*, p. 9505.

rare occurrence for any one to observe the Coronella taking its food, Dr. Opel gives an interesting account of a contest between one of his snakes and a Slow-worm (*Anguis fragilis*).

In the year 1857 I was in possession of such a number of snakes of various species that I was obliged to place my Coronella in the same cage with a blindworm, which had been there for some time already. The two appeared to be good friends, and took no particular notice of each other. Both passed into their winter sleep as the cold came on ; and, with the return of spring, again woke up and shared the cage in peace, coiled up together on the side where the sun's rays struck warmest. The blindworm ate freely of the earthworms offered to it, though all attempts failed to induce the Coronella to take any food. Small lizards placed near it were allowed to crawl away without notice, and even young mice were disregarded. One morning (May 9) I observed a great commotion in the cage. At this time the Coronella had not cast its skin, nor had it eaten anything for nearly nine months. The blindworm was striving to escape the fixed gaze of its companion, which was following it all over their prison. I placed some fresh water in the cage, and just at that instant the snake threw itself with irresistible force upon the blindworm, fixed its teeth into its head, and, flinging fold after fold of its body round its victim, held it in a vice-like grip, exactly after the manner of the giant serpents of the tropics. So tightly, indeed, did it embrace the unhappy blindworm, that the contents of the latter's intestinal canal were violently forced out, and scattered over the glass sides. Each desperate struggle of the blindworm was followed by a closer grasp on the part of the Coronella, which looked exactly like a roll of tobacco, through which the extreme end of the blindworm's tail protruded.

Of course this contest ended in favour of the snake; and the blindworm was slowly devoured This latter act occupied more than three hours; for the blindworm was a large one, measuring eleven

inches in length. After its meal, the Coronella re-
freshed himself with a bath, and seemed to take
great pleasure in the water, which doubtless served
to mitigate the great heat of its very rapid digestion.
So active is that, that Dr. Opel remarks : " I have
known the portion of the prey enclosed in the
stomach to be in a state of decomposition while the
other portion was still outside the jaws." When
bathing, the Coronella generally takes great care not
to immerse its head, except during very warm
weather, and then it will keep its head beneath the
surface for a quarter of an hour at a time.

In reference to its hybernation, Dr. Opel observed
that this was neither so long nor deep as in many
other reptiles. Nor did the Coronella bury itself in
the sand, but always stretched itself, and slept on the
surface. The vitality of this species appears, however, to be one of its leading characteristics, as may
be gathered from the following circumstance :—In
the autumn of 1858 Dr. Opel was obliged to be absent from home for five weeks. Having no one with
whom to intrust his pets, he packed them into a
vasculum which was attached to his knapsack, and
used by him for botanical specimens. In this they
travelled the whole time he was from home. They
did not appear to be in the least affected by the
close confinement to which they were subjected, but
returned to Dresden in safety and good health. For

its home, he says that it chooses in preference rocky
ground, overgrown with wood, secreting itself among
stones and thick moss.   Though nowhere so common
as the ringed snake, several localities are named in
Saxony in which this species is found.

   This reptile belongs not only to a different genus,
but also to a different tribe of Colubrines, to the
ringed snake.   Dr. Günther's brief description of
the species is:—"Scales in twenty-one rows, and
scales bifid; upper labials seven.   Brown.   Back
with two (sometimes confluent) series of irregularly
rounded   dark   spots.   Hinder   maxillary   tooth
smooth."*   In size and appearance it approximates
more to the viper than the snake, but, like the
latter, is perfectly harmless.   It never attains a
great length, the maximum being scarcely two feet.
"The head is but slightly distinct from the body;
the tail short and strong at the base; the eyes
small; the rostral plate presses much upon the
muzzle, and is of a triangular form, with its top
pointed; there are seven labial plates on the upper
lip on each side, the third and fourth of which touch
upon the eye; the scales of the body are smooth,
rhomboid, in nineteen longitudinal rows.   The plates
on the belly number from 160 to 164; those on the
under surface of the tail from sixty to sixty-four

---

* Dr. Günther's " Catalogue of Reptiles in British Museum."

pairs. The upper maxillary teeth are on the same line with the others, and longer. The upper parts are greenish-brown, with two parallel rows of black markings along the back, more distinct towards the head than in the hinder portion ; sometimes the spots on the back are small and few in number. The lower parts have a lighter ground-colour, but are often much darkened by black markings."* We have been more prolix in this description than otherwise we should have been ; but it is desirable that a minute detail of the chief features of this snake should be disseminated in order that it may be recognised and recorded, so as more fully to establish it as a truly indigenous species.

Of its disposition we at present know but little from experience. Mr. Bartlett says:—"I could not help observing that it is much more fierce than the common snake ; it bit me several times, but without any injury to the skin, in consequence of the shortness of its teeth." And this is confirmed by Dr. Opel, who says that, as a rule, it is irascible and ever ready to bite. In this, however, individuals vary. While the two he procured from Silesia never attempted to use their teeth, were particularly gentle, and suffered themselves to be taken up without making any effort to escape, others would bite at the

* *The Zoologist*, p. 6731.

F

finger on the slightest provocation, and hang on by their teeth, so that it took some little violence to remove them. He makes, however, one very consolatory observation upon this portion of his experience, to the effect that he never found the slightest inflammation, or other ill consequence, follow from the bite; and from numerous experiments he fully bears out opinions elsewhere expressed, that neither mammals nor birds are in the least degree affected by the bite of the *Coronella lævis*, or "Smooth Snake."

Viper or Adder.

PLATE V

THE VIPER, OR ADDER.

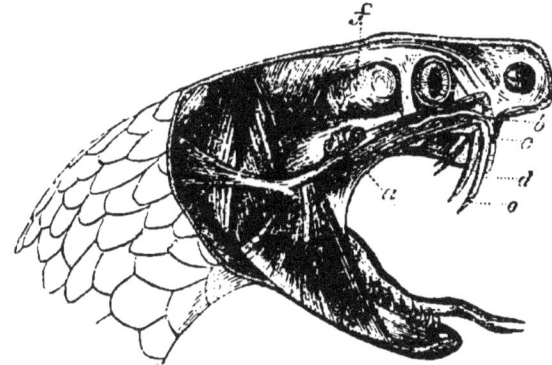

(*Pelias Berus.*)

FORTUNATELY for us, the present is the only venomous species which inhabits our island, and this is by no means equal to the venomous little reptiles of tropical countries in poisonous power. In Scotland, the Adder is common, whilst the Ringed Snake is but little known, and in England and Wales it is as abundant as any could wish. Of course it is unknown in Ireland. Whether Saint Patrick did or did not banish them, matters little ; it is, however, certain that in these latter times no snake-like reptile is to be found there. The Viper (Plate 5), or

F 2

Adder, for in some districts it is known by one name
and in some by the other, is not such a lover of
water as the snake, and may generally be found in
dry woods and heaths, in sandy banks, and similar
localities. It has been said that they are more than
usually common in the dry woods on the chalky soil
of Kent, and they certainly come nearly within the
sound of Bow bells, for they have been met with in
the little woods around Hampstead, Highgate, and
Hornsey.

It is a common belief that the venom of the
Viper, and other serpents, is almost innocuous in
winter, and its virulence is proportionate to the
heat of the weather, whether at home or abroad;
and that the snakes of tropical climes are more
deadly venomous than those of temperate countries,
on account of the greater heat. Recently Dr.
Guyon has set himself to investigate this subject,
especially whether the poison is innocuous in winter,
with the following results :—

Regarding its violence, he says there is a general
belief abroad that it is much more powerful in
summer than in winter; but this he does not
consider well authenticated, and quotes against it
the case of one Drake, an exhibitor of snakes, who,
having in the summer of 1827, at Rouen, handled a
rattlesnake which he took to be dead, while it was
only benumbed by the cold, was bitten by it and

died in the course of nine hours. From a considerable number of observations, Dr. Guyon concludes
that the intensity or power of the venom is less
owing to difference of season than to the length of
time it has been accumulating in the reservoir of
the reptile; and the greatest accumulation necessarily occurs during winter, because the animal is in
a torpid state and does not take any food during that
season.   So it was in the case of Drake, and so Dr.
Guyon found it in that of a horned viper which had
been given to him at the caravanserai of Sidi-
Makhlouf, Algeria.   This reptile had been put into
a bottle, which had since remained hermetically
closed.   It had been in there for six weeks, without
food and without air, and looked quite dead, since it
could not stir in the bottle, which it filled entirely.
And yet, on opening the bottle, the doctor found
the reptile perfectly sound, and saw it kill a large
fowl instantaneously with its sting.   Our author
quotes another case, that of a scorpion, that had
been kept in a bottle for a long time, and on being
released killed two sparrows in less than a minute,
and a pigeon in three hours.

A circumstance has come to our knowledge which
occurred in Warwickshire, of a boy that was bitten
by a viper during the winter :—

The 19th of January, 1864, was an unusually warm and sunny
day for the time of year. A boy, aged 11 years, started for a

walk to Kenilworth, about two miles from the village where he
lived. Under one part of the road flows a small brook. The boy
had his dog with him, and he wished to see whether he would
follow him across, as the stream was shallow, and there were
some large stones to step upon. The opposite bank is rather
steep, and there were several large pieces of wood and roots
projecting from it. One of his leggings was caught in the roots
and became unfastened, whereupon he sat down on one of the
stumps to refasten it and watch his dog in the water, but not for
more than two or three minutes. They then started off again,
and had not gone more than a quarter of a mile before the boy
felt a sharp pain in his wrist, On looking at his wrist he saw
plainly, and to his great horror, three little punctures and
places, as though a nettle had stung him. His first impression
was that he had been bitten by an adder, and he immediately
tried to bite out the piece of flesh. The first time he could not
manage it, and after two unsuccessful attempts, the third time
he bit out the flesh, sucked the wounded part, and at intervals
spitting out the poisoned blood.

Being fond of natural history, he remembered reading some of
the particulars about the bite of an adder, and was frequently in
the habit of expressing his fear of going where the grass was
long, lest he should meet with one. He went on a little farther,
but feeling faint and weak, and getting frightened about the
bite, he returned home. On arriving there he could hardly
speak, from excitement and the haste he had made ; but his first
words were : " Mother, I think I have been bitten by an adder:
though I did not see one, I feel and see on my wrist the hard
white swelling which always comes after a bite."

His mother immediately put his arm into very hot water, and
then applied a bread and oil poultice. When the doctor arrived,
he said all was done right, and the boy had saved his own life by
the courage and presence of mind he had shown in at once
biting out the piece of flesh. The poison, however, had swelled
up his arm through his veins as high as his shoulder ; but by the
next day this black streak of poison, apparent in the veins of
his arms, had completely disappeared, and in the course of a few
weeks the boy was perfectly restored to health.*

* Hardwick's *Science Gossip*, vol. i., p. 131.

The poison apparatus of the Viper (see woodcut at the head of this chapter) consists of the *gland* in which it is secreted, the *duct* or canal along which it travels, and the *fang* by means of which it is injected. The *gland* is placed at the side of the head (*a*), and consists of an assemblage of lobes. The substance is soft and yellow, with a spongy appearance. The *duct* or canal through which the poison is conveyed to the fang is a narrow cylindrical tube (*b*), swelling in the centre of its course into a kind of reservoir, and terminating at the base of the fang (*c*). This latter is a tooth in the form of a tube, much longer than the other teeth, and curved (*d*). It is placed in the upper jaw, one on each side of the mouth. Behind these are the similarly shaped, but smaller, fangs of replacement. On the outer surface of the fang, near the apex, is an elongated opening or slit (*e*), from which a canal passes through the hollow in the interior of the tooth and is united to the duct which communicates with the poison-gland. These fangs fall backwards, and lie concealed in a groove of the gum when not in use.

The following elaborate description of the mode by which the Viper wounds and envenoms its prey scarcely leaves anything to be desired:—" When a Viper is struck, it first coils itself up, leaving its head in the centre or at the summit of the coil, and drawn a little back, as if for the purpose of recon-

noitering. Speedily the animal uncoils itself like a
spring. Its body is then launched out with such
rapidity, that for a moment, the eye cannot follow
it. In this movement the Viper clears a space
nearly equal to its own length; but it never leaves
the ground, where it remains supported on its tail
or posterior portion of the body, ready to coil itself
up again and aim afresh a second blow, if the first
should fail. To do this the Viper distends its mouth,
draws back its fangs, arranges them in the right
direction, and then plunges them into its enemy by
a blow of the head or upper jaw: this done the
fangs are withdrawn. The lower jaw, which is
closed at the same moment, serves as a point of
resistance and favours the entrance of the poison-
fangs; but this assistance is very slight, and the
reptile acts by striking rather than biting. There
are times, however, when the Viper bites without
coiling itself up and then darting forth. This occurs,
for instance, when it meets with some small animal,
which it destroys at leisure and without anger, or
when it is seized by the tail or middle of the body,
in which case it turns round and plunges in its fangs.
As the teeth are buried in the tissues of the body
struck, the poison is driven down the canals which
pass through them by the action of the muscles
which close the mouth, and the injection takes place
with a force proportionate to the vigour and rage of

the reptile, and the supply of poison with which it is furnished."* In the bite there are two punctures corresponding to the poison-fangs.

It has been taken for granted that the bite of the Viper proves fatal in this country, without perhaps a knowledge of instances in which it so terminated. Professor Bell declares that he had never seen a case which terminated in death, nor had he been able to trace to an authentic source any of the numerous reports of such a termination which have at various times been confidently promulgated.† Nevertheless, so recently as the present summer (1865), a case was reported in the papers, in which a woman was bitten in Epping Forest, and shortly afterwards died : and we know that in France and other continental countries many instances are recorded. Bedard, in his lectures, relates a case of a young man in the neighbourhood of Angers, who, falling down in a meadow, was bitten by a viper in several places, and died in consequence in a few hours. Matthiole records an instance in which a countryman falling down in a meadow happened to divide one of these reptiles in the middle ; he seized the portion of the trunk to which the head was attached, in an awkward manner, and was in consequence bitten in the finger, and died from the effects of the wound. It should

* M. Moquin-Tandon's " Zoologie Médicale."
† Bell's " British Reptiles," p. 59.

be remembered, in connection with these instances, that the reptile which is regarded as the common viper in France is the *Asp*, and not the same species as that which occurs in Britain, whilst our Viper, not uncommon also, is called the Little Viper. The former of these is doubtless more venomous than the latter.

It may be of interest to record here the experiences of a sufferer from the bite of a viper, and the mode of treatment resorted to:—

The viper was very sluggish; but on touching it and endeavouring to take it by the neck, as I had done before, it struck at me, and bit my left forefinger. I immediately threw it down and stamped upon it, and sucked the place, cutting it with a knife, and putting ammonia to it. Meanwhile, my brother went for the doctor, and before he came I began to feel very faint, and inclined to vomit. When he came, he cut two incisions in the finger, and tied it up at the root, putting it into as hot water as I could bear. He also made me drink a quantity of ammonia and water, and then go to bed. The finger was very painful, and bled for a long time, and I was very feverish. The next day I got up, but was very weak, and there was a green mark all up from my finger, which was very large, to my left side, and all up my arm. The mark as far as the side was in a narrow line, but at the side it was in a large blotch. My finger was poulticed, and I had a cushion, dipped in water in which broken poppy-heads had been boiled, placed between my arm and side. The arm was much swelled, and I soon went to bed again. The next day I got up again, and from thence began rapidly to get well, and in a week had my arm out of a sling, and my finger almost well. I was perfectly well by the day fortnight on which the bite occurred. These facts are, I need hardly say, perfectly true, and the more remarkable because olive oil was not used at all.*

* *Science Gossip*, vol. i., p. 160.

The Viper is ovo-viviparous, that is, the young escape from the egg just before, or at the time of parturition, the membrane being very thin and easily ruptured. From ten to twenty young are produced at a birth, and these are as vigorous and pugnacious as their parents immediately on their entrance into the world, Before the event takes place which ushers so many little reptiles into society, the parent female, then in a more sluggish and inactive condition than at other times, may often be observed lying in the full blaze of sunshine, to acquire from that source the heat which is essential, and which her own cold blood fails to supply.

Gilbert White says : " On August 4th, 1775, we surprised a large viper, which seemed very heavy and bloated, as it lay in the grass basking in the sun. When we came to cut it up, we found that the abdomen was crowded with young, fifteen in number, the shortest of which measured full seven inches, and were about the size of full-grown earthworms. This little fry issued into the world with the true viper spirit about them, showing great alertness as soon as disengaged. They twisted and wriggled about, and set themselves up, and gaped very wide when touched with a stick, showing manifest tokens of menace and defiance, though as yet they had no manner of fangs that we could find, even

with the help of our glasses."* Again adverting to
this, lest it should be considered that he favoured
the popular notion that the viper swallows its young
on the advent of danger, he adds, "There was little
room to suppose that this brood had ever been in
the open air before, and that they were taken in for
refuge at the mouth of the dam, when she perceived
that danger was approaching, because then probably
we should have found them somewhere in the neck,
and not in the abdomen."

During the winter, vipers, snakes, and other
reptiles retire to some snug spot to hibernate. At
this period a company of vipers may sometimes be
found locked in each other's coils, in a hole at the
foot of a tree, or in some other out-of-the-way place.
Here they remain without food till the spring, when
they come out with the sunshine to enjoy themselves.   During the summer, the "sloughing"
process takes place, once or oftener, according to
circumstances; the skin becomes loosened, turned
back, and as the reptile glides amongst the grass its
old coat is slipped off and left behind. It is much
brighter in colour after this change than before.

Does the Viper swallow its young?—The belief
has a firm hold in the minds of many, that, on the
approach of danger, the young of the viper glide to

* " Natural History of Selborne," Let. xxxi.

their parent for protection, and that she opens her mouth, and, one by one, they pass down her throat, where they rest in security till the danger is past. To prove a negative is always a difficult task, but the effort to remove a prejudice must be even greater to be successful. Clergymen, naturalists, men of science and repute, in common with those who make no profession of learning, have combined in this belief, and to them we are indebted for many such accounts as the following :—" Walking in an orchard near Tyneham House, in Dorsetshire, I came upon an old adder basking in the sun, with her young around her; she was lying on some grass that had been long cut, and had become smooth and bleached by exposure to the weather. Alarmed by my approach, I distinctly saw the young ones run down their mother's throat. At that time I had never heard of the controversy respecting the fact, otherwise I should have been more anxious to have killed the adder, to further prove the case."* Nothing can well be more positive, clear, definite, and many would think *decisive*, than the foregoing ; yet so sceptical are some men on this subject, that they still dare to doubt whether there may not be some error in the observation. Let us advert to other witnesses, and evidence still more complete,

Rev. H. Bond, South Petherton, Somerset, in *Zoologist*, p. 7278.

and we do so with as earnest a desire for truth as
the witnesses themselves, and to know that the
debate is closed for ever.

J. H. Gurney, Esq., of Catton Hall, near Norwich,
well known as an ornithologist, and especially for
the splendid collection of Raptorial Birds in the
Norwich Museum, which has been obtained chiefly
through his instrumentality, in the year 1863, communicated to the *Zoologist* the following instance,
told to him by a person in whose accuracy he had
the fullest reliance. "John Galley saw a viper at
Swannington, in Norfolk, surrounded by several
young ones; the parent reptile perceiving itself to
be observed, opened its mouth, and one of the young
ones immediately crept down its throat; a second
followed, but after entering for about half its length,
wriggled out again, as though unable to accomplish
an entrance. Upon this Galley killed and opened
the viper, and found in the gullet, immediately
behind the jaws, the young one which he had seen
enter, and close behind that a recently swallowed
mouse. Galley was of opinion that the first young
viper which entered was unable to pass the mouse,
and that consequently there was not sufficient room
for the second young one, which endeavoured unsuc-
cessfully to follow in the wake of the first."*

* *The Zoologist*, p. 8856.

To this we may add another instance corroborative
and yet more conclusive, on the faith of a clergyman
with whose name and address we are furnished, and
in whose testimony we have the greatest confidence.
" Now, 'seeing is believing,' and I well remember
having seen in my boyhood—some thirty years ago
—an instance of the fact, the truth of which is
doubted, because resting merely on the testimony of
unscientific country people. Now, I have no pre-
tentions to science, but I vouch for the truth—above
referred to—of having, in my boyhood—when out
on a birds'-nesting expedition, in a southern county,
with some three or four companions—come suddenly
upon a viper sunning her young brood on an open
grassy spot in a broad hedge-row : hedge-rows were
common in those days. Immediately she saw us,
she began to hiss, and away went the young,
previously some feet from her, ' helter-skelter ' towards their mother; rushed into her mouth —
expanded to an immense width for so small a creature
—and down her throat, one over the other, while
you could say ' Jack Robinson.' The space where
she was recreating was some twenty feet square, so
that before she could beat to cover, we, boylike,
being armed with sticks, had beaten her to death.
This done, one of the party with his knife opened
the body, and out came again the little ones, all of
which we killed. I do not remember the exact

number, but my impression is that it was not more
than six or eight."* Another gentleman recently
communicated to *Science Gossip* the following
occurrence :—

> Some years since I was shooting in a wood, and came
> suddenly on a viper lying on a sunny bank. As soon as the
> viper caught sight of me, it began to hiss, and I distinctly
> saw several young ones, about three or four inches long, run
> up to the parent and vanish down its throat; and from the way
> in which the parent kept its mouth open, and the young ones
> glided into it, I should say they were accustomed to that sort
> of thing.†

We must not forget that some time since the
following occurrences were narrated in the *Zoologist*,
by the editor himself, and whilst they strengthen
the evidence of the Viper swallowing its young,
further serve to establish the fact of *viviparous*
reptiles being addicted to that habit. Both these
illustrations refer to the "Scaly Lizard," which,
like the Viper, brings forth its young alive. " My
late lamented friend, William Christy, jun., found
a fine specimen of the common Scaly Lizard with
two young ones; taking an interest in everything
relating to Natural History, he put them into a
small pocket vasculum to bring home, but when he
next opened the vasculum the young ones had disappeared, and the belly of the parent was greatly

* *Science Gossip*, p. 108.
† *Science Gossip*, p. 160.

distended : he concluded she had devoured her own offspring. At night the vasculum was laid on a table, and the lizard was therefore at rest ; in the morning the young ones had reappeared, and the mother was as lean as at first."

" Mr. Henry Doubleday, of Epping, supplies the following information :—' A person whose name is English, a good observer, and one, as it were brought up in Natural History under Mr. Doubleday's tuition, once happened to set his foot on a lizard in the forest, and while the lizard was thus held down by his foot, he distinctly saw three young ones run out of her mouth ; struck by such a phenomenon, he killed and opened the old one, and found two other young ones which had been injured when he trod on her.' In both these instances," Mr. Newman adds, " the narrators are of that class who do know what to observe, and how to observe it ; and the facts, whatever explanation they may admit, are not to be dismissed as the result of imagination or mistaken observation." *

We must confess that our own incredulity has been so staggered of late by these and similar instances, that we are by no means disposed to deny, because we cannot fully comprehend, the mystery of the process. It is admitted by some

* *The Zoologist*, p. 2269.

physiologists, if not by all, that there is no sound
physiological reason against such an occurrence;
and, until we are convinced by better arguments
than have hitherto been advanced, we are bound
to admit that in "our inmost hearts" there lurks a
belief that the maternal viper·has a knack of swallowing its young. Whether our scientific friends
consider us renegade from the true faith or not, we
will at least be true to ourselves,

Vipers were formerly held in some estimation as
a medicine; Pliny, Galen, and others extolled their
flesh for the cure of ulcers. Very recently, in the
French tariff, they were subject to a duty of four
shillings per pound. In Italy, a stew or jelly of
vipers is said to be regarded as a luxury.

The poison of vipers still appears to have some
reputation amongst medical men practising in the
East. Dr. Honinberger, late physician to the
Court at Lahore, seems to have had faith in it
for "rumbling in the bowels." He thus details the
manner in which he procured the virus, which
on one occasion was obtained from the Aspis
Naja :—

The man who brought the serpents to me, having wrapped
his hand in a cloth, took them by the back of the neck, and,
with a small stick, forced open the mouth, when, by means
of a pair of forceps, I held a small lump of sugar under the
tooth, above which is the bladder containing the poison ;
and, on his pressing the bladder with the stick, a drop of

limpid fluid fell through the tubular tooth on the sugar,
which I instantly deposited in a porcelain mortar, moistening
it with a few drops of spirit, and commenced trituration ; I
then put the powder into a small phial containing one drachm
of proof spirit, shaking them together—when it was fit for
use.   I kept it in a box, secluded from light, and before
administering it, shook it well up; one drop constituted a
dose.*

In concluding our account of this reptile, it may
not be out of place to indicate the chief features
by which the Viper may be distinguished from the
Snake.

It is the *Snake* that is harmless, and the *Viper*
that is venomous ; the latter being probably less
so in winter, and most dangerous in the hottest
weather, because then the secretion is more
rapid, induced by the greater activity of the
reptile.   The Snake is commonly the larger of
the two, and is found in the dampest situations,
generally in near proximity to water, in which it
delights to bask.   The Snake has large plates
or scales, upon its head, few in number; in the
Viper they are numerous and small.   The Snake
has no continuous line of a darker colour running
along its body, but is spotted all over ; the Viper
has a continuous line, zigzag and blotched, running
down its entire length.   The head in the Snake

* Dr. Honinberger's " Thirty-five years in the East," vol. ii.,
p. 230.

is more depressed and acutely pointed in front than in the Viper, which latter has a characteristic blotch something like the " death's head and thigh bones " of the "death's head moth," on the top of its cranium. Whether or not its venom is fatal, we would strongly advise our readers not to permit the Viper to make an experimental dart at their shins. It is better to indulge in a shudder when only a harmless snake crosses our path, than make the mistake of hugging a viper to our bosom.

Snakes or Vipers are not perhaps the easiest of all animals to be determined by a novice; at any rate they will require some little observation at first, until the eye is accustomed to see and recognise the differences whereby one species may be distinguished from another. There are nevertheless some points of difference more than usually distinct between the Viper and the Snake, which will be apparent on comparing the descriptions and figures. It may be premised that in depth and tone of colour there is considerable variation in some reptiles, and this is especially the case with the Viper, so that no reliance should be placed on that as a feature. In general colour it is often brownish or olive, but this may become nearly black, or it may be pallid grey, nearly white, or of a warm-red brown. The markings, however, are more permanent, a dark mark between the eyes, a spot on each side the hind part of the head, an obscure V, as though it bore the initial of its name on its crown, and a broad zigzag line down the whole length of its body and tail, apparently formed by the confluence of a series of dark lozenge-shaped spots, with irregular triangular spots on each side, are the chief features in the marking of this species. The scales, less visible without a closer examination, are also distinctive. Those on the top of the head are *not* large plates as in the innocuous snakes, but a greater number of smaller scales, three being larger than the rest. The scales of the back and sides are distinctly keeled, and disposed in eighteen series. Those of the under parts vary in their number, but are generally from one hundred and forty to one hundred and fifty, with about thirty-five pairs to the tail. It is seldom so large as the common snake, and the female, as amongst rapacious birds, is the largest.

Winged Scarabæus and Asps, from an Egyptian ornament.

AMPHIBIA, OR BATRACHIANS.

WE have already intimated the reasons wherefore
the Batrachians are now separated, more widely than
heretofore, in a scientific sense, from the true Reptiles, although they are still associated in the popular mind with them in almost indissoluble bonds.
This class includes those of the whole order Reptilia,
in its broadest sense, which at some period of their
lives inhabit the water, and are truly aquatic, and at
another are either wholly or chiefly terrestrial.
There are some very singular creatures in this group,
such as the Salamanders, to which such romantic
stories of their incombustibility pertain, and of whom
it is said, "If a salamander bites you, put on your
shroud." As late as 1789, a French Consul at
Rhodes, hearing a loud cry in his kitchen, rushed to
learn the cause, when his cook, in a horrible fright,
informed him that he had seen a "certain person-
age," who shall be nameless, in the fire. The Consul
affirms that he thereupon looked into the bright
fire, and saw a little animal with open mouth and
palpitating throat. He took up the tongs to secure

it, but, at first it scampered into a corner of the
chimney, lost a bit of its tail, then hid amongst some
hot ashes. It was ultimately secured, found to be a
little " lizard," was put into spirits, and sent to the
celebrated naturalist Buffon. Thus runs the story,
of which we must permit our readers to believe in
proportion to their credulity.

As the Batrachians or Amphibians, to which the
toads, frogs, and newts belong, differ from the Reptiles, already described, in their being at one portion
of their existence, entirely aquatic, and at other
periods entirely, or in part, dwellers on the land,
some account will be necessary of the transformations which they undergo. Professor Quatrefages
has well described these changes :—

In this group we meet with both complete and incomplete
metamorphoses; but here we find them marked by quite different features from those among insects. The changes do not
appear to take place suddenly, nor is there anything like tho
apparently torpid condition of the pupa. All the transforma
tions take place gradually, and as far as the external organs are
concerned, the development may be closely watched by the
observer.

The development of frogs presents another curious phenomenon. It is this : the young animal, after it has left the egg,
and before it has become a larva, is still in a semi-embryonic
condition. At this period the digestive tube and its appendages
are exceedingly rudimentary. The greater portion of the body
is filled by a large mass of yolk or vitellus, inclosed by the skin,
which has been formed for some time ; and it is at the expense
of this alimentary matter that the development proceeds.

The external characters are in keeping with the imperfect
condition of the animal at this period. The head is large, and

appears to be divided in two on the under surface, each half
being prolonged as a sort of process by which the animal
attaches itself to surrounding objects; as yet there are no
traces of either eyes, nostrils, respiratory or auditory organs;
and the belly, of an oblong form, is continued posteriorly as a
short tail bordered with a riband-like membrane. This primitive
condition, however, does not last long. About the fourth day
after birth, the head, which is now as long as the body, has
somewhat the appearance of a thimble; the mouth is provided
with a pair of soft lips; the nostrils, eyes, and auditory apparatus have made their appearance; the head is separated by a
deep groove from the belly, which has assumed a spherical form,
and from which spring a pair of opercula, clothed with little
branching gills; and the tail has grown so much that it is now
quite as large as the body. The mouth is very soon armed with
a horny beak, capable of dividing the vegetable food; the intestine, which is now very long, becomes more fully formed, and
assumes a spiral arrangement; the tail is elongated and widened,
and the little creature is then called a *tadpole.*

At this period, one of those alterations occurs which are so
intimately associated with the ideas we are endeavouring to
convey, that we must not pass them by in silence. Our larva
first breathed by its skin alone, and afterwards by a pair of little
branching gills attached to the opercula. About the seventh or
eighth day, however, the opercula are gradually soldered to the
abdomen, and the gills fade away and disappear. At the same
time a set of new and more complex branchia are developed, in
chambers situate on either side of the neck. The new gills are
arranged in tufts attached to a solid framework of four cartila-
ginous arches, and are about a hundred and twelve in number
for each side of the body. Here we see a rapid substitution of
one organ for another, though both discharge their functions in
the same manner, inasmuch as the respiration is just as aquatic
in character after the alteration as it was before it.

But the modifications of the respiratory apparatus do not
cease here. Before the tadpole can become a frog, it must do
away with these second gills and replace them by lungs; and at
the necessary time, a set of changes takes place analogous to
those we have already described. The vascular tufts are atro-
phied, and the lungs, which till now were solid and rudimentary,

open up and increase in size. The circulatory organs are corres-pondingly modified The calibre of the large branchial vessels is diminished, and the pulmonary trunks increase in number and diameter. Later on, the solid parts of the branchial appara-tus disappear also, the bones and cartilages being gradually re-absorbed. Eventually the alteration is fully accomplished, and there remains not the slightest trace of the former branchial apparatus. In this instance, not only has there been trans-formation and substitution, but an actual metamorphosis has occurred ; for the respiration, which was aquatic before, has become atmospheric, and, strictly speaking, the animal from having been a fish has been converted into a batrachian.

If we examine any particular apparatus, we shall find it also presenting many curious phenomena in the course of its develop-ment. We shall find that as the herbivorous habits give place to carnivorous ones, the digestive apparatus undergoes a change adapting it to the new form of diet. The mouth increases in size and gape ; the little beak-organs, or more correctly, the horny lips, are replaced by teeth, which are attached to the pala-tine arch, and not to the jaw ; the intestine, which before was long, and almost cylindrical, becomes shorter, and is inflated in certain portions of its length ; and the abdomen, which had been almost spherical, becomes thin and slender. The meta-morphosis, may now be seen in its entire extent, and more distinctly as regards the locomotive system than any other.

The tadpole at first exhibits no trace of either internal or external limbs. It swims about like a fish by the action of its tail, which is an extensive organ, longer and wider than the body, supported by a prolongation of the vertebral column, moved by powerful muscles, and supplied with large blood-vessels and numerous nervous branches. Beneath the skin and muscles of the anterior and posterior regions of the body, two little projections appear at a certain period. These are the limbs, and are at first attached to the adjacent structures by the nerves and blood-vessels which are supplied to them. These projections increase in size, their appendages appear in due course, and eventually the hip and shoulder bones are developed. As soon as these locomotive organs enter upon the discharge of their functions, the tail begins to disappear. Its skin, muscles, nerves, bones, and blood-vessels atrophy, and vanish from our sight. They have not faded away, they have not simply fallen off, they have not been cast off by a species of moulting, as in the case of insect larvæ. They have been got rid of by none of these methods; their substance has been re-absorbed, atom by atom; and hence, although it has ceased to exist, it is not the less alive on that account.

We see, then, that frogs undergo complete metamorphoses, not only in regard to their entire organism, but as to each set of apparatus, with the exception of the nervous system. The salamanders are not similarly situated. These maintain their external gills throughout the entire larval period, and never acquire internal branchiæ. When reaching the air-breathing condition, they skip as it were, one of the transformations which frogs undergo. The salamanders have also four legs in a perfect state; but then, in addition to these, they preserve the tail *

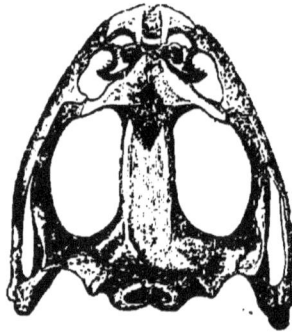

---

* "Metamorphosis of Man and the Lower Animals." Translated by Dr. Lawson. London : Hardwicke.

PLATE 17

1 Common Frog   2 Edible Frog

LITH. IN HOLLAND BY EMRIK & BINGER, H. LIEPNERS. LONDON

THE COMMON FROG.

(*Rana temporaria.*)

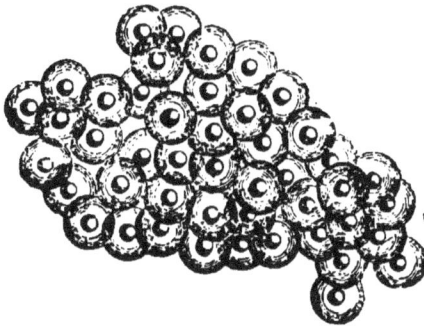

Eggs of the common frog.

Tɪıs Batrachian (Plate 6, Fig. 1.) is common
enough everywhere in Great Britain for us to dis-
pense with any very lengthened description. It
occurs all over Europe, from the south to the extreme
north, being abundant at North Cape, and not uncommon in Italy. In Ireland also the frog has
become naturalized, having been introduced about
the beginning of the eighteenth century.

Frogs spawn most commonly about the middle of
March. A large number of small round opaque

bodies, enclosed in a glairy mass, are deposited at the bottom of pools and ditches. Speedily, by the absortion of water, the envelope swells, and each ovum, like a little black dot, is enclosed within its sphere of gelatine. These masses of ova soon rise and float on the surface of the water, in which state they are known to all country people as "frog spawn."

At first the embryo is a small globular body, rather darker on one side than the other. In about four-and-twenty hours the sphere elongates, and in eight-and-forty hours the presence of a head and tail may be distinguished. Gradually the fin appears around the tail, and the rudiments of the *branchiæ*, in the form of tubercles, project from each side of the neck. Within about four days from the deposit of the ova, in a warm climate, such as Italy, the membrane is ruptured and the tadpoles become free; but in our own clime a much longer period is required, and the eggs are not usually hatched for a month after their deposition.

Soon after the emergence of the tadpoles from the egg, the *branchiæ*, or extended gills, attain their full development, when they gradually diminish in size, until at length they become withdrawn into the cavity prepared for their reception, and are closed by a fold of the skin. For some time the tadpoles continue to grow day by day, and increase in bulk without any material change in form. If carefully watched, however, little tubercles or buds will be observed in the course of time to make their appearance both towards the upper and lower portions, or rather forwards and backwards on the body ; these are the rudimentary legs.

As the latter grow and manifest themselves, the tail, which is in danger of becoming a useless appendage, is gradually absorbed, until the little " froggies," with their stumps of tails, cease to be " tadpoles," and merge into veritable little " frogs." It is very usual for persons living in towns to patronize tadpoles in their aquaria ; but so many are their misfortunes that a colony of town-bred frogs seldom gladdens the eyes of a Cockney. The tadpoles have a cannibal propensity to kill and eat each other, as their limbs begin to bud. " I placed in a large glass globe of water several tadpoles," says Mr. Bell, " more or less nearly approaching their final change, and I observed that almost as soon as one had acquired limbs it was found dead at the bottom of the water, and the remaining tadpoles feeding upon it. This took place with all of them successively excepting the last ; which lived on to complete its change, and for a considerable time afterwards." Whilst in the earlier stages, at least, of their existence, the green confervoid vegetable deposit of the tank appears to be the legitimate

food of the tadpole, seasoned perhaps with a small
quantity of minute animal life, the Frog is almost
entirely insectivorous. Its favourite food consists
of all kinds of minute insects, such as the little
green plant-lice, which are the pest of gardeners,
other larger insects, small slugs, and such forms of
animal life. This habit ought to procure for frogs,
not only the protection, but the fostering care of
gardeners and all cultivators of the soil. How much
less cause would they have to complain of insect
enemies, if they would but exercise more care in
the preservation and increase of toads and frogs,
and establish on their own domains a kind of " local
game law," instead of winking at the persecution, if
not really encouraging the extirpation, of their
best friends. It is strange that Nature should so
well provide for the " balance of power," and that
man should so pertinaciously endeavour to overturn
her work, by the wholesale destruction of insectivorous birds, or the persecution of harmless and
beneficent reptiles. Yet in the primeval forests,
where sparrow clubs are unknown, and insectivorous
animals dwell unmolested, no evidence can be traced
of any great mistake which Nature has made in the
preponderance of any given forms, nor can we trace
any manifest " bungle " which requires the wisdom
of " the lord of the creation " to put right.

Of all reptiles submitted to incarceration none

behave better in captivity than this, as the following testimony will prove :—

In a fern-case, about three feet by one foot in area, I kept two toads, a small frog, and a number of newts. The frog did not appear fond of the water ; the newts would not go in, and if thrown in, immediately crawled out again. The toads, on the contrary, appear to enjoy an occasional bath, remaining in the water, with the mouth and eyes above the surface, for several hours together. They lived, the one for about a year, the other for nearly two. The newts dropped off one by one, the last surviving for, I think, upwards of eighteen months. The frog lived for several months, and was a very interesting creature. When on the upper part of the fern-fronds, where he delighted to bask, he appeared of a distinctly greenish tint; but when on the soil, at the bottom, the hue changed to so decided a brown that it was difficult to find him. During the time that these creatures were among the ferns, I am not aware of having seen an *aphis*, whereas since their decease, the young fronds (especially those of the Polypods) are, during the summer months, infested with them.[*]

The voice of the frog is not generally regarded as particularly charming, and its vocal efforts are commonly called " croaking ; " but there is another kind of " croaking," uttered by animals of a much higher order in creation, which we regard as far less musical. A concert of frogs, heard remote from towns and railways, on a quiet evening, is not so inharmonious as " croaking " might lead us to suppose. There may be something in association, being ourselves of East Anglian birth, but we certainly like to hear an occasional " frog concert."

[*] Hardwicke's *Science Gossip*, vol. i., p. 86.

Mr. C. Darwin mentions a similar feeling which took possession of him at Rio de Janeiro :—

After the hot days, it was delicious to sit quietly in the garden, and watch the evening pass into night. Nature in these climes chooses her vocalists from more humble performers than in Europe. A small frog, of the genus *Hyla*, sits on a blade of grass, about an inch above the surface of the water, and sends forth a pleasing chirp: when several are together they sing in harmony on different notes.*

Probably the American frogs are really more musical than ours, as they have to compensate much for the loss of singing-birds, or at least melodious ones. The " Old Bushman " says that in Sweden a little frog emits during the pairing season a note like the ringing of bells, and as this sound proceeds from the depth of the water, it appears to come from a long distance, although the frog may be within a few fathoms.

During the winter these Batrachians, in common with others of their race, proceed to winter quarters, hiding themselves in holes of the ground, but more commonly buried in mud, congregated together in considerable companies. With the early spring they awake from their torpidity, break up their association, commence their career of love, and seek fitting localities for the deposition of ova and development of their young. In and about ditches

* Darwin's "Journal of Researches," p. 29,

and swamps, bog and fen, treacherous to human
feet,

By night or by day, were you there about,
You might see them creep in, or see them creep out.

Dr. Hermann Masius says of the Frog he may be
looked upon as a character; in popular stories, fairy
tales, and poesy, he plays no unimportant part. Full
of meaning is the myth of the Frogs of Latona;
also the fable of their election of a king. Æsop
related it to the Athenians, when Pisistratus had
usurped the government; and very lately its influence was put to the test by working it up into a
political drama. Aristophanes had already brought
the Frog people on the stage, just as two thousand
years later it appears to have furnished one of the
greatest German satirists with a welcome subject.
Unfortunately, of Fischart's "Froschgosch" the
name alone is preserved; another poem however,
which, in the heroic style of the Homerian epic,
sings the battle between the Frogs and Mice, affords
us a compensation. I mean the "Batrachomyoma-
chia," and the new version of it by Rollenhagen.
Appearing as it did towards the end of the sixteenth
century, "Froschmäusler" was long, and very justly,
a favourite book of Protestant Germany. In 1787,
when Prussian troops marched into insurrectionary
Holland, a third frog epic appeared, as if to show
how inexhaustible the subject was. Thus has this

H

race of animals won for itself an inalienable place
in poesy; and, beginning with fable and with riddles,
runs through the whole of song.

The Koran relates of Mohammed that he knew
how to appreciate these despised animals, for that
he caused them to be respected for having saved
Abraham from a fiery death. When the Chaldeans
had thrown the patriarch into the flames, in order
to kill him, frogs came compassionately to the
rescue, spat water into the fire, and extinguished it.

We have not adverted to two very extraordinary
phenomena attributed to these animals. If the
frequency with which accounts of these events
illustrate the columns of our newspaper press may
be accepted as proof of their verity, then there are
no mysteries of Batrachian life so true as " showers
of frogs," and live frogs or toads found embedded in
coal, limestone, granite, &c., immured for ages, or
since times contemporaneous with the patriarch
Noah. We know it is not philosophic to deny,
because we cannot explain, any of the marvellous
things recorded of sea-serpents, buried toads, or
frog-showers; nevertheless we will venture to *doubt*,
whilst believing that future generations will, one
day, set these matters at rest, beyond the possibility
of doubt. It would be folly to deny that there are
still very many estimable people in existence who
believe in showers of frogs : perhaps a still larger

number with faith in imprisoned reptiles, such as
the toad in the coal of the Great Exhibition, or the
frog celebrated in the *Gainsborough News* in the
following paragraph :—

A few days ago a very fine live frog was discovere l imbedded
in a large block of stone, at the Lady Lee stone-quarries, near
Worksop, Nottinghamshire, now occupied by Mr. J. Ellis. The
block was eleven feet below the surface, and the frog on being
liberated, jumped about quite cheerfully, and on being placed in
a pond of water, swam with great dexterity. It is supposed the
prisoner must have been confined from one to two thousand
years. The block of stone had the impression of the frog very
distinctly marked where it had lain for such a long period.*

We had well-nigh forgotten to associate with
these the " true and particular account " of

Sir Froggy who would a wooing go,
Whether his mother would let him or no.

The climbing powers of frogs attracted the attention of the Rev. C. A. Johns, of Winchester, and
in a letter recently published, he gives the following
instances :—

Three several instances proving that they can and do climb,
have fallen under my own notice. These are already recorded
in print. A fourth came under my notice on the 27th of October
last (1863). I was digging for pupæ at the base of a large
willow-tree in the valley of the Itchen, near Winchester, with
some young friends, when one of the party exclaimed, " Look
at this frog climbing up the tree ! " I quickly ran round to the
other side of the tree, and saw, not one only, but five or six
young frogs, from one to two feet from the ground, climbing up
the rugged bark, and using their front and hind feet just as a

* *Retford and Gainsborough News*, March 11th, 1865.

sailor employs his hands and feet when ascending the rigging
of a ship. One which I did not myself see was discovered at a
height of five feet from the ground in the act of descending.
It had been alarmed probably at our intrusion, and had fallen
to the ground before I reached the spot; but I had no reason to
doubt the accuracy of the statement, for two or three members
of my party pointed to the exact spot from which it had fallen;
and if a frog can climb two feet, there is no reason why it
sho·ld not climb twenty, or more.*

Mr. Henry Reeks afterwards affirmed that he had
found frogs frequently in apparently inaccessible
places, such as the tops of pollard willows, in the
vicinity of streams, to which they could not have
attained save by climbing. The same gentleman
also alludes to the toad as an adept in climbing,
having often found them in the nests of small birds,
in hedges.† Other observers have borne out the
above testimony that frogs *do* climb.

The common frog has its head nearly triangular;
the teeth are minute, and arranged in a single row
in the upper jaw, with an irregular row across the
palate but none in the lower jaw. The tongue is
lobed at the tip, and folds back upon itself when not
in use. The fore feet have the third toe the
longest, and the second the shortest. The hind legs
are more than half as long again as the body, the
toes webbed, the fourth being one-third longer than
the third and fifth. The skin is a little wrinkled on

* Rev. C. A. Johns, in *The Zoologist*, p. 8861.
† *Zoologist*, p. 8927.

the thighs, but is elsewhere smooth. The upper parts are brownish, or yellowish-brown, sometimes very dark, and spotted with black, and the legs assuming the form of bands. There is also an elongated patch of brown or black on the temples, with an indistinct paler line down each side of the back. Length about three inches.

Tadpole, natural size, and enlarged, in its early stage.

THE EDIBLE FROG.

(*Rana esculenta.*)

In naming the geographical distribution of the
edible frog (Plate 6, Fig. 2), the author of "The
Quadrupeds and Reptiles of Europe" says, "This
frog is found all over Europe, except in the British
isles; throughout the north of Asia to Japan, and
in Egypt." We venture to take exception to this
verdict, and affirm that the edible frog *is* found in
the British isles, as well as the rest of Europe ; but
whether it be truly a native, is another question,
the affirmation of which we do not intend to
be so positive in giving. There exists amongst
some naturalists too great a desire, or perhaps
rather a habit, to regard the British Flora and
Fauna as distinct from those of Europe, instead of

looking upon them as forming portions, isolated though they may be by geographical position, of the great continental whole. As an island, we possess, not only idiosyncrasies but also animals, insects, or plants which are somewhat peculiar ; yet because we are surrounded by the sea *now*, that is no reason why we should regard our " tight little island " as a world of our own, and forget our relationships with the old continent from which we " seceded " many long years ago.

The following is Mr. Berney's account of his introduction of the edible frog into this country.

I went to Paris in 1837, and brought home two hundred edible frogs and a great quantity of spawn. These were deposited in the ditches and in the meadows at Morton, in some ponds at Hockering, and some were placed in the fens at Foulden, near Stoke Ferry. They did not like the meadows, and left them for ponds. In 1841, I imported another lot from Brussels. In 1842, I brought over from St. Omer thirteen hundred, in large hampers, made like slave ships, with plenty of tiers ; these were moveable, and were covered with water-lily leaves, stitched on to them, that the frogs might be comfortable and feel at home. These were dispersed about in the above-mentioned places, and many hundreds were put into the fens at Foulden and in the neighbourhood.*

In 1853 Mr. A. Newton and his brother were driving along the road between Thetford and Scoulton, in Norfolk, and hearing an unusual noise, stopped to ascertain the cause. His brother alighted

* *The Zoologist.* p. 6539.

and entered the meadow whence the sound proceeded, but soon returned and informed Mr. Newton
that the sound proceeded from a pond, and that the
musicians were edible frogs.

Of course (writes the latter gentleman) I went immediately to
satisfy myself, and there, sure enough, were the frogs—some
swimming to and fro in the water—some sitting on the aquatic
plants, with which the pond was choked, and these last were exceedingly noisy, puffing out their faucial sacs, like so many
dwellers in the cave of Æolus. After observing them for a little
while, we tried to obtain some specimens, but herein fortune
favoured the frogs. We had no aggressive weapons beyond a
walking-stick and an umbrella, and they were wary to a degree,
and exceedingly active. However, by persevering we became
possessed of four individuals, three, I regret to say, dead, and
one, an indiscreet youth, whom we found rambling about the
grass alive. We retired with our spoils, the deceased were de-
cently embalmed, and are now in the Norfolk and Norwich
Museum.

He afterwards relates how he made his discovery
known to Mr. J. H. Gurney, and learnt from him
that many years before Mr. George Berney had imported a number of these reptiles alive, and liberated
them in this neighbourhood. Application to Mr.
Berney educed the foregoing account of his introduction of edible frogs into this country.

How far the facts above related, adds Mr. Newton, will serve
to answer the inquiry propounded nearly twelve years since by
Mr. J. Wolley (*Zoologist*, p. 1821), "Is the edible frog a true
native of Britain?" I do not presume to say, but I will merely
draw attention to one point, namely, that as appears from Mr.
Berney's letter, upwards of 1,500 edible frogs, besides spawn,
had been by him alone imported into the East of England, some
of them six years, and all more than a whole year, before Mr. C.

Thurnall's discovery of the species in September, 1843, at Foulmire Fen, in Cambridgeshire.*

The above evidence would appear very conclusive,
so far as evidence of that kind could go, had the
publication of Mr. Newton's remarks not elicited a
reply from Mr. Thomas Bell, which was published in
a succeeding number of the same journal, in which
his remarks appeared. Mr. Bell says:—

My father, who was a native of Cambridgeshire, has often described to me, as long ago as I can recollect, the peculiarly loud,
and somewhat musical, sound uttered by the frogs of Whaddon
and Foulmire, which procured for them the name of "Whaddon
Organs." My father was always of opinion that they were of a
different species from the common frog, and this opinion of his,
formed nearly a century ago, was confirmed by Mr. Thurnall's
discovery that the frogs of Foulmire are of the species *Rana
esculenta.* †

The croak of the edible frog, as alluded to already,
is much louder and more musical than that of the
common species.

"Let me," says a German writer, "recall our
summer nights of Northern Germany. When on
the wide plain all life is asleep, and the lonesome
disquieting groan of the moor-frog is all that is
heard sounding from afar, like a summons from the
nether world, on a sudden the frog in the pond begins to raise his voice. It is an agreeable tenor. He
summonses to horary prayers : in a large circle round

* *The Zoologist*, p. 393.
† *Ib.*, p. 6565.

about him sit the synagogue; when presently a
deeper voice and evidently of one advanced in years,
chimes in; a third joins the chant, and the recitative
begins. A little while, and a pause ensues; then
the precentor sings again alone, some long-drawn
responses follow, when suddenly a hurly-burly, that
thrills through every fibre, bursts forth in the air. It
lasts some minutes, until single solos, in a minor
key, disengage themselves from the scattered tones,
which soon break forth again in a stormy chorus.
Thus does their music last on throughout the whole
night, and may be heard for many miles. Yet this,
it would seem, is but gentle music, when compared
to the uproar which hums in the ear of the traveller
when, on the shores of the Volga and the Caspian
Sea, the frogs in myriads celebrate their marriage
festivals. The bacchic rejoicings of these orgies ab-
sorb everything; all is grown froggish; it is as
though the very earth were shaking with the rest,
unable to resist the inextinguishable laughter."

The proportion of European frogs found in Great
Britain is small. Out of the nine species inhabiting
the continent we have but two. The little painted
frog (*Discoglossus pictus*) is confined to the extreme south on the shores of the Mediterranean.
The bell frog (*Alytes obstetricans*) is smaller still,
and has a wider distribution, being found principally
in the central countries, and is not uncommon in

France. The brown frog (*Pelobates fuscus*), which gives out a strong odour of garlic when touched, is apparently less common, but is found in France, Belgium and a few other localities. The rough-headed frog (*Pelobates cultripes*) has only been observed in Spain and the south of France. The red-bellied frog (*Bombinator igneus*) is found over the temperate regions of Europe, in France, Germany, Switzerland and Russia, passing most of its time in water. Lastly the common tree frog (*Hyla viridis*), with the extremities of its toes expanded into a kind of cushion or disk. Except during the spawning season, it is an inhabitant of trees and shrubs. This species is found nearly everywhere in Europe, except the British Islands. It is a small light green, elegant species, with a loud and musical croak. The seven species just enumerated, with the two more fully described as British, constitute the frog fauna of Europe. Of toads there are but three, and two of these belong to our islands.

The name of "edible frog" suggests a gastronomic idea not much in favour with Englishmen. To cook and eat frogs was long since regarded, on this side of the Channel, as the privilege only of Frenchman. Recently we have been relaxing in our prejudices, and both Frenchmen and frogs, seen through a less distorting medium, are being better understood and appreciated. "*Fricassée de grenouilles*," which we

should called "preserved frogs'-legs," are regularly
sold at some of the West-end provision warehouses,
packed in air-tight canisters, in the same manner as
many other mysterious modern comestibles. It is
three or four years since we learnt that "French
frogs" had introduced themselves into British commerce, and we resolved upon testing their edible
qualities. One of the canisters referred to was
purchased, and by dint of labour opened, and what
seemed to be little strips of boiled chicken floated
in thin melted butter. We looked at them, and
paused—smelt of them and paused again;—at length,
with marvellous courage, as we thought, tasted them.
At first there seemed some little irresolution on the
part of the internal genii, whether they ought not to
rebel against the admission of the intruder. But, as
the taste was in their favour, down went the frogs.
After all, the flavour was very agreeable, if one
could only forget for a while their origin and name,
and fancy them little stewed rabbits, which they
most resemble.

Mr. F. Buckland, when in Paris, on one occasion,
resolved upon a banquet of frogs.

I went (he says) to the large market in the Faubourg St. Ger-
main, and inquired for frogs. I was referred to a stately-looking
dame at a fish-stall, who produced a box nearly full of them,
huddling and crawling about, and occasionally croaking, as
though aware of the fate to which they were destined. The
price fixed was two a penny, and having ordered the dish to be

prepared, the *dame de la halle* dived her hand in among them and having secured her victim by the hind legs, she severed him in twain with a sharp knife ; the legs, minus skin, still struggling were placed on a dish ; and the head, with the fore-legs affixed, retained life and motion, and performed such motions that the operation became painful to look at. These legs were afterwards cooked at the *restaurateur's*, being served up fried in bread-crumbs, as larks are in England ; and most excellent eating they were, tasting more like the delicate flesh of the rabbit than any-thing else I can think of. I afterwards tried a dish of the common English frog, but his flesh is not so white nor so tender as that of his French brother.

Frogs, and sometimes toads, are, more exten-sively eaten than some of us would imagine. In China, at New York, on the banks of the Seine and the Amazon, in the West Indies and the East, in the Phillippines and the Antilles, by both barbarous and civilized races, frogs or toads are regarded as delicacies. Professor Duméril used to warn his pupils that the dealers, in collecting frogs, often met with toads, and never rejected them, but cutting off the hind quarters and skinning them, mixed all together, toads and frogs, and sent them to Paris to be eaten. The Chinese have a peculiar taste with regard to frogs, and economize portions which the Parisians reject. Mr Fortune says :—

They are brought to market in tubs and baskets, and the vendor employs himself in skinning them as he sits making sales. He is extremely expert at this part of his business. He takes up the frog in his left hand, and with a knife which he holds in his right, chops off the fore part of its head. The skin is then drawn back over the body, and down to the feet, which are chopped off and thrown away. The poor frog, still alive but headless, skinless, and feetless, is then thrown into another tub, and the operation is repeated on the rest in the same way. Every now and then the artist lays down his knife, and takes up his scales to weigh these animals for his customers, and make his sales. Everything in this civilised country, whether it be gold or silver, geese or frogs, is sold by weight,

One of the remembrances of our school-days is of one or two of the big boys who were accustomed to astonish their juniors by the singular exhibition of putting little frogs in their mouths, and some-times swallowing them. In Cheshire, we are told, it is still the practise to catch little frogs and place them alive in the mouths of children suffering from the *thrush*. Of course, this is considered an infal-lible cure, and in some eastern counties the disorder itself is known as "the frog." The victims in these cases being the common frog, that species would appear to have some little claim also to be regarded as edible; though, on the other hand, if swallowed alive, how can they be said to have been eaten? *N'importe!* a dead rabbit is better than a live frog, as an esculent, provided it is not too "high."

Frogs and toads had formerly some reputation in medicine, either wholly or in part; but all belief in their efficacy is now vested in a few obscure matrons, who prescribe for the rustic population in some out of the way villages in agricultural districts. *Sic transit gloria mundi.*

The principal features by which the edible frog

may be distinguished from the common frog are :— the absence in the former of the conspicuous dark patch, which in the latter extends from the eye to the shoulder—the vocal sacs or bladders at the angles of the mouth in the edible frog, which are distended while croaking, and which are absent in the common frog—and the light line which runs down the back of the former but which is not seen in the latter. To these may be added the more distinct and beautiful markings in the edible frog, its louder note, and generally larger size. In their food, habits, and habitats, there is probably no great difference between them, except that the edible frog appears to be more exclusively aquatic. Messrs. Duméril and Bibron, the authors of a large French work on Reptiles, say that:—

It inhabits indiscriminately running or still waters, the borders of rivers, rivulets, or streams, lakes or ponds, salt or fresh marshes, or even ditches and simple pools of water. Sometimes they are seen on the leaves of water-lilies, or on the herbage of the banks, where they love to bask in the warm sunshine; but at the slightest noise they strike into the water, and do not again expose themselves until certain that all danger is past.

In the edible frog the toes are cylindrical, and a little swollen at the tips; the webs of the toes are slightly notched, and do not reach the extreme tips; the fourth toe of the hind foot is one-fourth longer than the third and fifth; the nostrils half-way between the corner of the eye and the tip of the

muzzle; the head is triangular; the teeth on the
palate are in a line exactly between the nasal
openings; the tongue is broad, lobed, and covered on
the surface with scattered warts. On the upper
surface of the body are a number of rather indistinct,
scattered warts or folds, but the skin on the belly is
smooth. The male is furnished with a bladder at
the angle of the gape on each side, which, when
distended is as large as a small cherry. The colour is
exceedingly variable, generally of a greenish tint,
sometimes of a rich chestnut red. Length above
three inches.

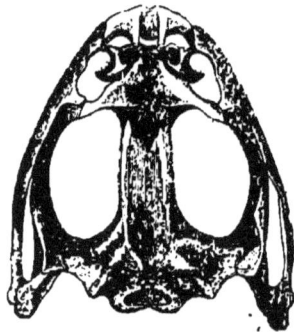

1 & 2 Natterjacks, 3 Common Toad

PLATE VII

THE COMMON TOAD.

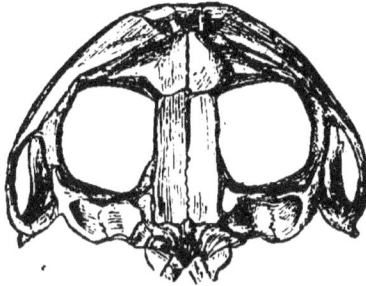

Skull of Common Toad (enlarged).

ONLY a small proportion of the European species of frogs is found in Britain; whilst of three toads, two are indigenous to these islands. Some have even doubted whether two of these species (*B. viridis* and *B. calamita*)) are not varieties of the same. If this be the case, all the European toads are British. We think, however, that the two just named are really distinct, and that the green toad (*Bufo viridis*) is, therefore, a stranger and an alien.

The common toad (Plate 7, Fig. 3) is found all over Europe, from Sweden and Russia to Greece and Italy, Ireland excepted.

I

Pennant commences his chapter on this animal
in the following words :—

> The most deformed and hideous of all animals ; the body
> broad, the back flat, and covered with a pimply dusky hide : the
> belly large, swagging and swelling out ; the legs short, and its
> pace laboured and crawling ; its retreat gloomy and filthy : in
> short its general appearance is such as to strike one with disgust
> and horror ; yet we have been told by those who have resolution
> to view it with attention, that its eyes are fine. To this it seems
> that Shakespeare alludes, when he makes his Juliet remark :—
>
>   " Some say the lark and loathed toad change eyes."
>
> As if they would have been better bestowed on so charming a
> songster than on this raucous reptile.*

Pennant, however, only gave expression to the
feeling which was common, and almost universal, at
the time he wrote. Toads were looked upon with
dread and disgust ; and even now many people, not
only illiterate, but educated, would describe it in
equally prejudiced terms. There is nothing very
prepossessing, perhaps, in a casual glance at a toad
—nothing to recommend it very strongly to a lady
as a drawing-room pet. How often have we heard
the exclamation of an anxious mother to her child,
" Don't go near that toad; it will spit at you! "
And how long and earnestly might we plead with
such a mother before we could convince her that
the toad would do her child no harm !

* Pennant's " British Zoology," vol. iii., p. 14.

Before adverting more particularly to the habits of this animal, we will endeavour to give a brief epitome of what is known of the secretions of the toad, which were in part, perhaps, the basis of the belief in its injurious character.

Dr. Davy thinks that the principal use of this poison is to defend the reptile against the attacks of carnivorous animals. He also remarks that, as it contains an inflammable substance, it may be excrementitious; it may serve to carry off a portion of carbon from the blood, and thus be auxiliary to the functions of the lungs. In support of this idea, the author observes that he finds each of the pulmonary arteries of the toad divided into two branches, one of which goes to the lungs, and the other to the cutis, ramifying most abundantly where the largest follicles are situated, and where there is a large venous plexus, seeming to indicate that the subcutaneous distribution of the second branch of the pulmonary artery may further aid the office of the lungs, by bringing the blood to the surface to be acted upon by the air.*

The viscid exudation from the skin in this kind of Batrachian is regarded as a kind of poison. It appears to answer as a defence to an animal which has no other means of defending itself; but, as there are no provisions for inoculation of the fluid, it cannot be employed as a means of attack. According to M. Moquin Tandon, it is a thick, viscid, milky fluid, with a slight yellow tint, and poisonous odour. It has a disagreeable caustic, bitter taste; becomes solid on exposure to air, and

* "Philos. Trans. for 1826," pt. ii., p. 127.

assumes the form of scales when placed on glass.*
M. Pelletier affirms that its acrid properties are due
to the presence of an acid. MM. Gratiolet and
Cloez performed with it some experiments on birds,
such as linnets and finches, which were inoculated
with the fluid, and died in about six minutes. They
were not convulsed, but opened their beaks, and
staggered as if in a state of drunkenness. In a short
time they closed their eyes as if falling to sleep,
and fell down dead. The same gentleman also
ascertained that when a small quantity of the fluid
was introduced beneath the skin of such mammalia
as the dog or goat, it caused death in less than an
hour.

M. Vulpian repeated these and similar experiments, both with the common toad and the natterjack. He inoculated dogs and guinea-pigs, and
found that they died in from half an hour to an
hour and a half. This fluid acts also as a poison on
frogs, and generally kills them in the course of an
hour; it is sufficient to apply it externally upon
their backs; but upon toads themselves it has no
influence.

It is said that in certain countries the Indians
hunt after several species of toads with pointed
sticks. They transfix the animals with these sticks,

* Moquin Tandon's " Medical Zoology," p. 287.

and when they have collected a considerable quantity of them, they place them before a large fire, but at a sufficient distance to prevent their being roasted. The heat excites the cutaneous secretion, which is collected by the Indians as it is discharged from the pustules, for the purpose of poisoning their arrows. The humour secreted in the follicles of the triton, or great water newt, has similar properties, but is less virulent.

In a recent number of a French journal, an instance is recorded of the virus of the toad entering the blood of a child, and causing death :—

A young lad, ten years of age, named Louis P——, whose parents are small tradespeople in the Faubourg Saint-Antoine, was playing with some of his companions near Bercy, not far from a building in the course of demolition. This boy, who was of a delicate constitution, had a slight abrasion of the skin of the right hand. Having seen a lizard crawl into a hole in an old wall, he put in his hand, but instead of the lizard he drew out an enormous toad, which he immediately threw on the ground. The skin of the toad is covered with large tubercles, formed by an aggregation of small pustules, open at their summit. A milky liquid, of a yellowish-white colour, very thick, and of a fetid odour, escapes from these tubercles when the animal is irritated. Whilst the lad had the animal in his hand, this liquid, which is a violent poison, was introduced through the wound in his hand into the blood. He was soon after seized with vertigo, vomitings, and faintings, and was carried to the house of his parents, who called in a doctor immediately; but already the malady had made such progress that, in spite of the most energetic means employed, the patient soon died.*

* *Petit Journal*, 29th March, 1865.

The venomous character of the toad seems to have been a firm belief of the ancients; but they had, at the same time, very erroneous opinions of the " toad's envenomed juice," assigned to it, a power and an action far different to what it possesses. Ælian regarded the toad as capable of conveying death in its look and breath; and Juvenal records of the Roman dames that they infused the venom of the toad in wine, to form a draught whereby to rid themselves of their husbands. Our own Shakespeare used it as a compound of the witches' caldron in " Macbeth ":—

> Toad, that under coldest stone.
> Days and nights hast thirty-one
> Sweltered venom sleeping got,
> Boil thou first i' th' charmed pot.

Even Pennant, with all his repugnancy to the toad, could not be induced to favour the popular belief in its poisonous character.

We shall now return (he writes) to the notion of its being a poisonous animal, and deliver as our opinion, that its excessive deformity, joined to the faculty it has of emitting a juice from its pimples, and a dusky liquid from its hind parts, is the foundation of the report. That it has any noxious qualities we have been unable to bring proofs in the smallest degree satisfactory, though we have heard many strange relations on that point. On the contrary, we know several of our friends who have taken them in their naked hands, and held them long, without receiving the least injury, It is also well known that quacks have eaten them, and have besides squeezed their juices into a glass and drank them with impunity. In a word, we may consider the

toad as an animal that has neither good nor harm in it, that
being a defenceless creature, nature has furnished it, instead of
arms, with a most disgusting deformity, that strikes into almost
every being capable of annoying it, a strong repugnancy to
meddle with so hideous and threatening an appearance.

The toad is very easily domesticated, and when
under confinement, or partially so, soon becomes
emboldened and apparently attached to those who
cater for its appetite. A gentleman who caught a
young toad, and brought it to the great metropolis
to reside with him in town, gives the following particulars of his pet :—

He would occasionally absent himself for weeks, so that I
ceased to be alarmed for his welfare, even though I might not
have caught sight of him for a month. In this manner we went
on, leaving Toady to take his holidays as he pleased, until the
spring of last year. During one of his temporary vacations, I
was watching the movements of some small insects, and it
appeared that my pet was watching them also, for on their
approaching within reach of his tongue that organ was instantaneously thrust forward, and the insect disappeared. Thus,
while losing sight of a new acquaintance, I became aware of the
presence of an old friend. I also derived fresh satisfaction in
his choice of food, and mode of taking it. Thenceforward I
became diligent in supplying him with the same kind of food, so
that he soon lost all appearance of shyness, would come out of
his hiding-place regularly, day by day, until late in November,
1863, when he again disappeared as the frost set in. At this
time the weather was very severe for so early a period: the
aquarium was frozen, the fish were killed, the glass was broken,
and all its contents became a solid mass, plants, animals, and
everything embedded, as it were, in a large transparent crystal.
Again, I was agreeably surprised, one beautiful spring day in the
early part of April, to observe my old friend moving about, as if
to inform us that he had returned again from his unknown place

of retreat. He had again grown fatter and uglier than he was in the autumn, his skin was blacker and coarser, and dark spots covered the whole body. Yet his eye seemed more brilliant and thoughtful. He came direct to the same spot on which I had fed him when last we met, more than four months previously. He was supplied with what we term "garden-hogs," wood-lice, worms, and the lively little black ant. None of these would he touch, if dead, or did not show unmistakable signs of active life. Then would he fix his calculating eye, until the object came within reach of his tongue ; this he would dart at them, and in an instant the object was gone. When satisfied, he would return again to some quiet nook, out of sight. This summer being long and dry, I have had some difficulty in providing him with his necessary food. One day I placed him in a large hole at the bottom of the garden, where I collected the sweepings and rubbish, and he literally became a "toad-in-a-hole." This was some fifty yards from the house, and I left him to shift for himself amongst the insect life of the rubbish. I afterwards sought him to convey him back to his old neighbourhood around the house, but he was nowhere to be found, and this time I gave him up for lost. Four days after, what was my surprise, whilst seated at supper, to see Toady come tumbling heels over head down the step into the room, on a visit to his old friends. The most remarkable feature in this last freak is, the circuitous route he must have taken before he arrived, and the obstacles he must have encountered in his way.*

Toads, as well as frogs, are insectivorous ; and it is curious to observe how, by means of their folded tongue, they rapidly catch and appropriate any small insects which may come in their way. Large beetles and earthworms occasion them much more trouble, sometimes annoyance.

Mr. Holland, in narrating his experiences with

* *Science Gossip*, vol. i., p. 12.

toads, gives an excellent illustration of their mode of
feeding :—

My toads, two in number, had lived for a year or two in a
hothouse which was devoted to the growth of pineapples. They
were, I think, first placed there purposely by the gardener, who
found them very useful in destroying insects. I used very frequently to visit the place and amuse myself with feeding the
toads with worms, and with watching their habits. The heat of
the place, which was considerable, did not seem to inconvenience
them in the least, for they were remarkably active, and of a large
size ; but at the same time they seemed greatly to enjoy the
artificial showers when the plants were syringed, and would come
out from their hiding-places to be rained upon. They usually
remained amongst the pineapple plants, which grew on a bed
raised some four feet from the ground, where they sat under the
long leaves; but when the place was watered, they would not
unfrequently jump down and lie upon the cool, wet tiles of the
floor, spreading themselves out as flat as possible. How they
climbed up to the pine-bed again I cannot say, for I never saw
them do it.

They evinced very little shyness, taking worms readily when
offered to them. When feeding, their actions were very curious.
Upon placing a worm about three inches from a toad, it would
instantly fix its attention upon it. Then its whole appearance was
changed. Instead of the dull, lethargic-looking animal that the
toad generally appears, it was all vivacity ; the body was instantly thrown somewhat back, and the head bent a little downwards, its bright eye riveted upon its prey ; and though the toad
was perfectly still, as long as the worm remained motionless or
nearly so, yet its attitude and its eager gaze were full of life
and animation. Directly the worm made any active movement,
the toad would dart forward, open its mouth from ear to ear,
and seize it, generally about the middle. A curious scene now
took place: mouth and feet went to work in good earnest ;
the worm was gulped down by a series of spasmodic jerks, trying
to make its escape every time the mouth was opened, the toad
thrusting it back all the time, and forcing it down its throat
by the aid of its fore-feet. Altogether it was rather a disgusting

sight, and gave one the idea that the toad is an uncommonly greedy animal.

Having got the worm down was by no means a reason that it would stay there, for I have sometimes seen a worm rather larger than usual make its way up again ; however, the feet would immediately go to work a second time, and the toad would at length remain the undisputed possessor of its own dinner.

Frequently I used to cheat the toads by moving a small twig before them.   They would seize it directly, imagining it to be a worm, and would regard it with stupid astonishment when they discovered their mistake.

I never had the good fortune to see my friends take beetles or other small prey ; but these, the gardener told me, were never seized with the mouth, but were caught with unerring aim upon the point of the long tongue.*

There is a curious twitching movement of the hind toes observable in the toad whilst watching an insect which he purposes making its prey.   This movement has been noticed to us by several persons who have kept toads in confinement, or closely watched their habits.

Mr. G. Guyon, of Ventnor, has kindly communicated to us some additional particulars of the behaviour of these Batrachians under confinement. There are two or three features in his letter so curious, and the whole so interesting, that we think no apology is needed for introducing it entire.

My toads are sufficiently tame to sit quietly on the hand while carried to the window, and there snap up the flies which they are held within reach of, and in this way I often cleared my sit-

* Hardwicke's *Science Gossip*, vol. i., p. 62.

ting-room of these troublesome insects during last summer
Indeed the notion occurred to me of constructing a sort of cage
of wire-work, and suspending it in the centre of the room, as is
done with the so-called " Fly-catcher," believing that the enclosed toad would be willing to make himself useful by appro-
priating the flies that settled on the wires, while visitors of
limited zoological attainments would be puzzled by the *strange
bird* thus hung up. Some similar arrangement attached to a
window would materially reduce the number of insects on the
panes. It has been observed that they refuse to seize anything
that is not in motion, and it may be added that they will attack
anything that *is*, provided it is not too large to be swallowed
It would appear as if their sight, or perhaps judgment, was very
defective, as the form or colour of an object seems of no import-
ance, if only it moves. It need not in the least resemble any
living creature, any small object in motion is regarded as suitable prey. If the end of a pencil, or anything similar, is drawn
along the side of the vase which the toads inhabit, they are
nearly sure to strike at it; and not learning wisdom by failure,
they will go at it again and again. One or two small tortoises
shared the same vase till the post proved too much for their
constitutions, and if one of these put out its head with the
intention of taking a slow " constitutional," the movement was
pretty sure to attract the attention of the nearest toad, who
would place himself in a convenient position, intently watching
his shelly companion, until the latter moved again, when dab
would go the toad's tongue upon its head, which the tortoise
would quickly draw back into the shell, not appearing to relish
the unceremonious salute. This absurd scene, which was of
frequent occurrence, was plainly owing to the toad mistaking
the head of the tortoise for some insect small enough to be
swallowed. Although the toads refuse to strike at a motionless
insect, they may be easily deceived, and induced to attack the
still object if motion is communicated to themselves, no doubt
thinking it is the object that moves, as a child in a railway
train fancies the houses and trees are running past. In this
way I have often, when holding them to the window, made them
strike at flies that were quite stationary.

On one occasion, while carrying the toad about, I noticed a
fly near the top of the door, and, presenting the reptile to it, was

surprised to find that though the usual snap was made, the insect never moved. On close examination I found that, being near-sighted, I had mistaken a fly that had been smashed in closing the door for a living specimen, and the toad had evidently made the same mistake. If they do not succeed when they strike at an insect, they will try again and again, in no way discouraged by failure.

In collecting flies for their benefit, a glass tube, closed at one end with a cork, has been found very handy. Flies at rest are easily caught by a rapid sweep of the hand, and if the open end of the tube is then thrust into the hand, the fly will quickly make its way into it, and thus one or two dozen flies can be readily introduced into the tube, as they will always collect at that end which is pointed to the light. If this tube is then placed in the toad vivarium, the insects will crawl out one by one, and as they proceed along it the toads will repeatedly strike at them, not deterred by the hard glass that balks their attempts. The swallow of the toad must be capacious, as insects of large size go down at one gulp. The cockroach of our kitchens, commonly called the "black-beetle," is a substantial insect ; but the largest specimen disappears as quickly as a house fly, though not unfrequently one of the antennæ is seen projecting cigar-fashion from the toad's mouth for a minute or two. Hard beetles, if of any size, appear more difficult to dispose of. I have seen a large Otiorhynchus *reproduced* with a dreadful grimace the next minute after it had been swallowed ; but a second attempt was more successful and the poor insect was seen no more in the land of the living.

The manner in which a toad manages to get rid of his old skin has been thus minutely described by an eye-witness :—

About the middle of July I found a toad on a hill of melons, and, not wanting him to leave, I hoed around him; he appeared sluggish, and not inclined to move. Presently I observed him pressing his elbows hard against his sides, and rubbing downwards. He appeared so singular, that I watched to see what he was up to. After a few smart rubs, his skin began to burst open, straight along his back. Now, said I, old fellow, you have done it; but he appeared to be unconcerned, and kept on rubbing until he had worked all his skin into folds on his sides and hips ; then, grasping one hind leg with both his hands, he hauled off one leg of his pants the same as anybody would, then stripped the other hind leg in the same way. He then took this cast-off cuticle forward, between his fore legs, into his mouth, and swallowed it ; then, by raising and lowering his head, swallowing as his head came down, he stripped off the skin underneath, until it came to his fore legs, and then grasping one of these with the opposite hand, by considerable pulling, stripped off the skin ; changing hands, he stripped the other, and, by a slight motion of the head, and all the while swallow. ing, he drew it from the neck and swallowed the whole. The operation seemed an agreeable one, and occupied but a short time.*

Professor Bell, in his " British Reptiles," gives a similar account to the above, upon the faith of his own observations. As the toad generally retires into his private apartments to undress himself, and dispose of his old clothes, opportunities for observing the process do not often occur.

As we never kept toads in confinement long together, for the purpose of observing their traits of character, we have been obliged, as our readers will have observed, to draw pretty freely from the experiences of our friends. This we have preferred doing in their own language, at the risk of all imputation of " scissors and paste," rather than rob those of honour " to whom honour is due."

* *New York Independent*, Dec. 29, 1859.

This creature is far more terrestrial in its habits than the frog, yet its ova are deposited and developed in water ; and the first six months of its career it is as aquatic as a fish. The eggs are arranged in long double chains, and not deposited in a mass, as is the case with the frog. There appears to be very little difference between them in their early stages. The ova are deposited two or three weeks later ; the tadpoles are similar but darker, pass through the like stages, and, in the autumn, having attained their legs and lost their tails, they venture upon the land, as miniature toads, and commence their terrestrial life. In this state they crawl about in search of their prey, neither running with the natterjack, nor leaping with the frog, but less vivacious than either, and more persecuted than both.

We have alluded to the incarceration of frogs in blocks of granite, &c., and, out of courtesy to the toad, which deserves as much at our hands, subjoin from the *Leeds Mercury,* an account there given of a truly patriarchal toad, at least if the assumptions of its historian are true :—

During the excavations which are being carried out under the superintendence of Mr. James Yeal, of Dyke House Quay, in connection with the Hartlepool Waterworks, the workmen on Friday morning found a toad embedded in a block of magnesian limestone, at a depth of twenty-five feet from the surface of the earth, and eight feet from any spring-water vein. The block of stone had been cut by a wedge, and was being reduced by the workmen, when a pick split open the cavity in which the toad had been incarcerated. The cavity was no larger than its body, and presented the appearance of being a cast of it. The toad's eyes shone with unusual brilliancy, and it was full of vivacity on its liberation. It appeared, when first discovered, desirous to perform the process of respiration, but evidently experienced some difficulty, and the only sign of success consisted of a "barking" noise, which it continues invariably to make at present on being touched. The toad is in the possession of Mr. S. Horner, the president of the Natural History Society, and continues in as lively a state as when found. On a minute examination, its mouth is found to be completely closed, and the barking noise it makes proceeds from its nostrils. The claws of its fore feet are turned inwards, and its hind ones are of extraordinary length, and unlike the present English toad. The Rev. R. Taylor, incumbent of St. Hilda's Church, Hartlepool, who is an eminent local geologist, gives it as his opinion, that the animal must be at least 6,000 years old. This wonderful toad is to be placed in its primary habitation, and will be added to the collection in the Hartlepool Museum. The toad when first released was of a pale colour and not readily distinguished from the stone, but shortly after its colour grew darker until it became a fine olive-brown.

Professor Bell devoted some attention to this question, and, not only him but other naturalists, without being convinced that such incarcerations for immense periods of time are proven. Mr. Edward Newman, of *The Zoologist*, we believe has, more than once or twice, inquired into the particulars of such accounts, and invariably found a flaw, and such a flaw as to prevent his coming to the conclusion to which such accounts tend. Whilst he and they are too good students of nature to deny,

or strive to mystify, what they do not comprehend,
they look too closely at facts, and are too chary of
deductions, without sound foundation, to accept
mere affirmations in lieu of proof. All we can say
is—"It may be so, and it may *not.*" The latter
alternative seems the most probable.

The "toad-stone" or "toad's jewel" was one of
the superstitions of a superstitious age, alluded to
in the lines—

> Sweet are the uses of adversity,
> Which, like the toad, ugly and venomous,
> Wears yet a precious jewel in his head.

—it having been supposed that in the head of the
toad was to be found a wonderful stone, which was
strong in curative power and magical virtue.

The most singular supposed association of toads
with some kinds of fungi, whence the latter have
been said to derive their name of "toadstools,"
evidently arose in a mistaken and foolish application of a new sense to an old compound, after the
original meaning and derivation had been forgotten.
This word is clearly derived from the German
"tod" and "stuhl," meaning "death-stool," in
reference to the poisonous nature of some, and the
supposed dangerous character of others, of the
stalked fungi, and, despite our woodcut, has nothing
whatever to do with toads, as commonly supposed.

In Norfolk those fungi, which in other localities are commonly called " toadstools," are called " toads'-caps," according to some orthography, but as we have heard it, " toad-skeps." Any one who may converse with farm labourers, or their children, about any of the Agarics, except the common mushroom and the Horse mushroom, will hear no other name for them than " toad-skeps." The etymology of this cognomen is to us very obscure, although conversant with the local dialect. The large wicker baskets, holding a bushel, and which are extensively used in East Norfolk, are there called " skeps"; but what connection there is between a toad and a bushel basket is to us as much a problem as the association of a toad and a side-pocket, another East Anglian allusion equally classical and inexplicable.

The body of the toad is broad, thick, very much swollen ; the head large, with the crown much flattened ; muzzle obtuse and rounded ; gape very wide ; no teeth either on the jaws or the palate ; the tongue entire, not notched ; over each eye a slight porous protuberance, a larger one on each side behind the ears; on the fore feet the third toe is the longest ; the first and second are equal, and longer than the fourth ; the hind legs scarcely longer than the body, with five toes, and the rudiment of a sixth, webbed for half their length ; the

K

fourth much the longest ; the third a little longer than the fifth ; the skin both above and below is covered with warts and pimples of various sizes ; these are largest on the back, but more crowded on the belly.

Length of the body about three and a half inches. Often larger in some parts of Europe.

THE NATTERJACK.

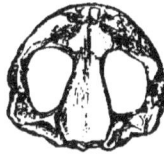

(*Bufo calamita*, Laur.)

THE Natterjack (Plate 7, Fig. 1 and 2), is less common or widely distributed in this country than its
congener the common toad. Yet it is far from uncommon in many localities, and may be regarded
rather as local than rare. In places where it is
found at all, it is often plentiful. The earliest
author who mentions this animal as a British native
is Pennant, in the third volume of his " British
Zoology." He says it had been found " on Putney
Common, and near Reverby Abbey, Lincolnshire,
where it was called the natterjack." It is to be
met with in several localities around London, as at
Blackheath, Deptford, Cobham, and Wisley. In
Cambridgeshire, near Gamlingay; in Norfolk; at

Ormesby, near Yarmouth ; and one or two stations
in the neighbourhood of Lynn and Norwich, and in
Suffolk, near Southwold. In Scotland it is recorded,
on the authority of Sir William Jardine, as occurring " in a marsh on the coasts of the Solway Firth,
almost brackish, and within a hundred yards of
spring-tide high-water mark. It lies between the
village of Carse and Sotherness Point, where I have
found them," he adds, " for six or seven miles along
the coast. They are very abundant." In Ireland,
Dr. Carrington verifies that they are to be found at
Ross Bay, and a correspondent of the *Field*, at
Roscrea, writes*:—

Mr. Tate sent me from England three dozen natterjacks.
I sent two to the Zoological Gardens, Dublin, and gave the
others their liberty about the place. We occasionally meet
with some of them, and they walk about rapidly, and can climb
anything in their way, in a most extraordinary manner, remind-
ing one of the movements of a lizard. I kept four of the
"natterjacks" for a few days. They lived upon worms and
slugs, and whenever I uncovered them, they immediately
concealed themselves amongst the damp moss given them for a
bed, and feigned death.

In form this species is very similar to the toad,
but may be easily distinguished by its difference
in colour. It is of an olive tint, darker on the
flanks, and with a definite pale yellowish stripe,
or line, running down the back. The under parts

* *The Field*, June 24th, 1865.

are yellowish with black spots, and dark bands
occur on the legs. The warts with which the upper
portion of the body is studded are of a reddish
brown. In its habits it is far less sluggish than the
common toad, sometimes indulging in an extempore
run, and altogether is far more attractive. Mr.
W. R. Tate, from whom we have received specimens,
has, for some time, kept them in conjunction with
other reptiles. He says : "They always go in pairs,
and seem to be more delicate than the common
species, as mine scarcely ever enter the water in
cold weather, which the latter frequently do. I find
them most commonly on sunny days, where a pond
has nearly dried up. Mine are now tame enough to
eat out of my hand. Their food consists of worms and
insects, which they catch by their tongues in the
same way as the other species. Their croak is
hoarser than that of the toad. A person inhabiting
a disused semaphore, on a heath in Surrey, says
that they do great mischief in his garden by
digging their holes in the seed beds. These holes
are dug straight for a few inches, and then there is a
passage at right angles to the perpendicular one, in
which the creature lies. The men call them
' Goldenbacks.' "

In the eastern counties, where we are told that
this species is sufficiently common to be recognised
by the country people as distinct, it is called the

" Walking Toad." It is a matter of history that
superstitions and old wives' fables are by no means
extinct in Norfolk, and we learn of one connected
with the " Walking Toad," which is extant in the
neighbourhood of King's Lynn. One of these toads
is to be obtained and buried in an ant's nest, where
it is to be left for some time. When the flesh is
all cleared off by the insects, and the skeleton is
quite clean, the shoulder bones are to be taken off
and thrown into a running stream. One of these
bones will float with the current, while the other
will float against it. The latter bone must be
secured, and, if kept as a talisman, will confer on
its possessor supernatural power.

The name of " natterjack " is evidently a corrup-
tion of the German. In that country they appear
to have been first known. We think that *natter*
was probably derived from *nieder*, Anglo-Saxon
naedre, " nether " or " lower," from the creeping
habit of the " adder," to which it belonged, under
the form of *eddre ;* and " Jack " from *jager*, " one
who runs," a very applicable term for such a running
reptile as the natterjack. Moreover, words com-
pounded of *nieder* have the signification of some
place or object lying low, and *jager* or *jagd* in
such a combination would be well applied to this
Batrachian.

Pennant notices of it that " several are found

commonly together, and, like others of the genus, they appear in the evenings;" but Mr. Tate has given a fuller account of the nocturnal habits of this species than we have seen recorded elsewhere:—

After dark, or at least after the rising of the moon, I was returning home across Wisley Heath, and when near a pond something ran quickly across the path. I took it up, and saw by its bright vertebral stripe, showing clearly in the moonlight, that it was a natterjack. I therefore commenced looking round the pond, and caught no less than fifty-seven of them. The noise they were making was very great; their croak being hoarse, and one continued note, instead of, as in the common toad and frog, a succession of short notes. The natterjack showed more sense than the toads, by leaving off croaking, and squatting close to the ground to escape observation whenever I approached one of their haunts, while the toads kept croaking and hopping. I found them always in shallow water (in which they can sit with their heads out), and, as their name implies, among reeds very often. I see now why their eyes are so much brighter by night than by day, as they are evidently nocturnal in their habits; but until this time I have always caught them on hot sunny days going about the heath in pairs.*

The "mephitic toad" of Shaw's "Zoology" appears to be the present species[2], for his description is that of an "olive toad, spotted with brown, with reddish warts, and sulphur-coloured dorsal line." Such being the case, we must take exception to some portion of his remarks:—

* W. R. Tate, in *Science Gossip*, vol. i., p. 111.

In its pace it differs from the rest of the toad tribe, running nearly in the manner of a mouse, with the body and legs somewhat raised. It is chiefly a nocturnal animal, lying hid by day in the cavities of walls, rocks, &c. The male and female perfectly resemble each other. They breed in June: so speedy is the evolution of the ova that the tadpoles liberate themselves from the spawn in the space of five or six days. This happens about the middle of June; and about the end of August the hind legs appear, which, in a certain space, are succeeded by the fore legs, and by September and October the animals appear in their complete form.

Roesel informs us that this species is known in some parts of Germany by the name of *roerhling*, or reed frog, from its frequenting in spring-time such places as are overgrown with reeds, where it utters a strong and singular note or croak. When handled or teased, it diffuses an intolerable odour, resembling that of the smoke of gunpowder, but stronger; this proceeds from a whitish acrimonious fluid, which it occasionally exudes from its pores. The smell in some degree resembles that of orpiment or arsenic in a state of evaporation; and sometimes the animal can ejaculate this fluid to the distance of three or four feet, which, if it happen to fall on any part of the room where the creature is kept, will, according to Roesel, be perceived two months afterwards.*

As far as this country is concerned, the foregoing is a great exaggeration of the odour emitted by this toad. Lord Clermont remarks: "When excited, it emits from the skin a strong sulphury odour." At present we have not experienced a very *strong* sulphury odour, much less any approximation to Shaw's description. There is, however, a very appreciable odour under certain conditions of excitement, &c.

* Shaw's "Zoology," vol. iii., pt. i., p. 149.

Mr. Hepworth, of Wakefield, has made some
curious calculations and investigations on the pro-
lific nature of frogs and toads, as well as contributed
useful observations on their enemies whilst in the
larval state.* Mr. Couch having recorded that on
one occasion he had drawn out and measured one of
the strings of ova deposited by the natterjack, and
had found it at least one hundred feet in length,
Mr. Hepworth calculates that as these strings are
double, and allowing eight ova to the linear inch,
the number of eggs deposited by one female toad of
this species would reach not less than nineteen
thousand. He afterwards makes an independent
calculation on the common toad : "I have taken
four inches as the average diameter of these masses
(of ova). Now, as there are eight eggs in one
linear inch, and six strings laid side by side fill the
same space, we shall have for one cubic inch 288
germs, and in a globular mass of four inches diameter 9,650, or rather more than half the number
obtained from Mr. Couch's measurement in the case
of the natterjack." This, he argues, is nearly the
same in effect, " when we consider that there are
two rows in the spawn of the natterjack, while in
the common toad there is only one." As the calculation is based upon the space occupied by a single

* *The Naturalist*, vol. i., pp. 24, 73, &c. Huddersfield, 1865.

ovum, whence the number contained in so many cubic inches is calculated, we cannot comprehend how *that* number is influenced by the ova lying in pairs or singly. Further, the observation that in one species the line is single, and in the other double, arises probably from regarding the alternate ova in the one chain as a single zigzag line, and not as a double chain with the ova alternately arranged.

After adverting to the uses of the tadpoles in the economy of nature, the same gentleman enquires, "What becomes of these myriads of tadpoles?" and sets about to furnish an answer :

Few of those vast swarms that blacken the waters in spring with their dusky forms ever reach the perfect frog. Their enemies are many, their means of defence few. They become the prey of larger or more warlike animals than themselves. These constant attacks greatly thin their numbers. Thus by the time they are fit to leave the water, they are, though still somewhat numerous, much less so than at an earlier period of their existence. But having left the waters, they are still exposed to great dangers. They are greedily devoured by the snake, weasel, polecat, and by nearly every species of water-fowl.

Amongst the creatures who feed largely upon the toad and frog in its larval state are enumerated the larvæ of the dragon-flies (*Libellula*), which are exceedingly destructive; the larvæ and imago of the great water-beetle (*Dytiscus marginalis*); the boat-flies (*Notonectidæ*); the various species of newts, and several fish, such as the bearded loach (*Gobitus barbatula*), and the stickleback (*Gasterosteus aculeatus*). Finally, he concludes that not one in a thousand of the young frogs which emerge from the egg in spring ever reach their winter quarters.

Were it not for the unbounded fertility of the frog and toad, they would be totally exterminated in one year by the unceasing attacks of their numerous terrestrial and aquatic foes. Should this fertility be checked by any cause whatever, these creatures, like their giant prototypes of the Mesozoic and Cainozoic ages, would soon be known only by their remains.

Its general appearance is similar to that of the last species, but the eyes are more projecting, with the eyelids very much elevated above the crown ; porous protuberance behind the eyes not so large ; toes on the fore feet more nearly equal; the third notwithstanding a little longer than the others; first and second not shorter than the fourth ; hind legs not so long as the body; the toes on these feet much less palmated than in the common toad ; the sixth toe scarcely at all developed ; skin similarly covered with warts and pimples. Above of a yellowish-brown or olivaceous, clouded here and there with darker shades ; a line of bright yellow along the middle of the back ; warts and pimples, especially the porous protuberance behind the eyes reddish ;

beneath whitish, often spotted with black ; legs
marked with transverse black bands.*

Length 2¾ inches ; hind leg 2¼ inches.

* Jenyns' "Manual of British Vertebrate Animals," p. 302.

PLATE VIII

Great Warty Newt ×

LITH. IN HOLLAND ST LONDON M NOEL 21 HERNERB ST LONDON

GREAT WATER NEWT.

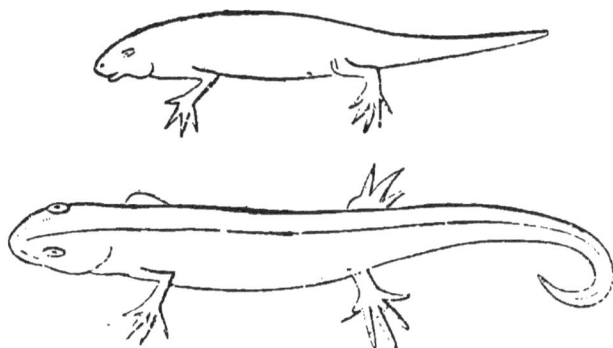

Great Water Newt, at the end of the first and second years.

(*Triton cristatus.*)

THE Great Water Newt, or Common Warty Newt (Plate 8), is only a small creature of its kind after all, since it seldom exceeds a length of six inches. Though not so common as the smooth newt, it is by no means a rare British species, and is generally to be found in ponds and large ditches, where it subsists upon water insects and small animals, occasionally making a meal of its cousins the little eft or smooth newt. On the Continent this triton is the most common species, extending in its geographical distribution from Italy to Sweden; is plentiful all through Italy and Switzerland, and is not uncommon in the south of France, Belgium, and Carniola.

Any one who has paid the slightest attention to the British newts will at least have noticed this rough-skinned species and the common smooth newt. It is now upwards of twelve years since a very patient and earnest observer made pets of the British tritons, and studied closely their habits and changes through a period of five years. Since we cannot lay claim to any such close and continued observation, the principal facts of our history of this species will be derived from this source, which is the most complete and authentic account we possess.

To commence with the egg, we learn that the ova begin to be deposited as early as the beginning of April, and continue to be deposited until the first or second week in July. These ova are carefully enclosed, singly, in the folds caused by the bending together of the leaves of certain aquatic plants. Those most favourable for this purpose have been found to be Water Speedwell (*Veronica anagallis*) and long grasses. If the leaf be too pliable and soft, it opens and exposes the ovum so that it perishes; if too rigid, the triton is unable to bend or break the fibres so as to fold it conveniently.

If a plant with long leaves be thrown into a pool where there are tritons, for only a single night during the breeding season, it will be found on the following morning to have a number of its leaves folded, and within each fold an ovum.*

When it is first deposited, the ovum or egg is of a globular shape, with a little white yolk in its centre, floating in a watery fluid, and surrounded by a delicate but firm transparent shell or capsule. This latter is covered on the outside with a gelatinous adhesive substance, which assists in securing the ovum to the leaf within which it is folded. If the egg becomes exposed too early to the full influence of the water, it is addled or rendered sterile. Within about fourteen days they become so large as to force the folds of the leaf apart, and expose themselves to the water, which at that period exerts no deleterious influence. From a series of experiments instituted by Mr. Higginbottom, it is clear that constant counter-currents are going on, of water passing into and out of the cell of the ova, through their transparent walls. This is doubtless necessary at the present stage for the perfect development of the eggs. At the end of three weeks the embryo is fully formed, moves freely within its envelope, and shortly escapes. External circumstances, especially temperature, exert considerable influence upon the growth of the embryo, and either hasten or retard

* Higginbottom, in " Ann. Nat. Hist.," 2nd ser. XII., p. 371.

its development. When the tadpole leaves the
ovum it swims away freely, and either attaches itself
to the sides, or falls to the bottom of the vessel. It
soon commences to feed voraciously, is not at all
particular about its diet, and will devour the tadpoles of the smooth newt, with no compunction on
account of their near relationship. " I have seen
the warty triton," says Mr. Higginbottom,* " in its
branchial state, with three of the smaller species in
its stomach at one time." The legs appear to be
very tardy in their development, and until they
possess sufficient strength to support the creature
on land it continues to inhabit the water in its fish-
like state. It is not until three months after their
exclusion from the egg that the triton-tadpoles give
any evidence of their quadrupedal tendencies. Even
at this period the hind legs are exceedingly delicate,
though the fore legs are developed in as many weeks.
When the legs are all fully formed the branchiæ are
absorbed, the gills are closed, and the triton emerges
upon dry land to enter upon its terrestrial existence.
This generally occurs about the middle of September. It may be noted that tritons which emerge
from the ova too late in the season to develop their
legs before the cold weather sets in make no pro-

* " On the Influence of Physical Agents on the Development
of the Triton."—*Philosophical Transactions*, 1850, part ii., p. 431.

gress during the winter, and do not attain a terrestrial state until the ensuing spring. Under ordinary and favourable circumstances, five or six months are necessary for their passage through their metamorphosis.

There were, prior to Mr. Higginbottom's paper becoming known, two puzzles in the history and habits of the tritons. Why were they so commonly collected from water, and yet, soon after being transferred to water in confinement, make every effort to escape from that element? And why were so many varieties and sizes, crested or uncrested, found, and yet so nearly allied as to appear but as forms of the same species? Answers to both these queries are found in the terrestrial habit and slow development of the reptile through three whole years after its change from the tadpole stage. "The triton does not commonly return to the water until the expiration of the third year, when it is so far advanced towards maturity as to be able to reproduce its kind." We have here a singular instance of protracted development, of a slow growth through four successive summers, and a cessation of growth through four inglorious winters. At length the full period of efthood is arrived at, when, the crested male having attained the dimensions of a perfect gentleman-newt, may take to himself a spouse, or the more soberly attired lady-newt lend a

L

willing ear or eye to the advances of the other
sex.

At the close of the first year the length of the
triton is about two inches, of the second year three
inches, of the third year four inches, and at the expiration of the fourth year it has reached from five
to six inches, thus adding annually about an inch to
its length.

In its earliest terrestrial state the skin is but
slightly rough, and through the first and second
years very little change takes place in this respect ;
but at the close of the third year the skin becomes manifestly rougher, and attains the per-
fection of its warted appearance in the fourth
year.

During the first and second years there is scarcely
any difference in the sexes ; but during the breeding
season, in the third year, the male aspires to an
incipient crest, the tail expands, and a permanent
silvery stripe appears along either side. The average
weight of the triton at the close of each of the first
four years is thus given by the observer already
quoted :—At the end of the first year 25 grains, of
the second year 54 grains, of the third year 75 grains,
and of the fourth year 134 grains. Hence, we learn,
that whilst the weight is more than doubled at the
close of the second year, but little more than one-
third is added during the third year, whilst in the

fourth year three-fourths of the prior year's weight becomes superadded.

At the close of March the *perfect* triton again resumes its predilection for the water, and returns to an aquatic life. At this period it becomes voracious, the male is full crested, with which appendage the female, generally the largest, is not adorned. After a little more than three months of this second life in the water, the crest of the male diminishes to a mere ridge, the female has deposited her ova, and both return again to a terrestrial existence until the following and succeeding years. Thus are the phases of their existence passed, and there is little reason to wonder at the fact that the same species should have been regarded as distinct, in some of these stages when the history of their career was unknown, and their metamorphosis little understood.

Hitherto we have not specially referred to the period of hybernation. Prior to this there is a general meeting of all ages, which occurs at no other period of the year, and takes place about the month of September.

In the summer the tritons are found in abundance in old brickyards. The brickmakers, who are constantly disturbing the water and removing the clay, and who occasionally clear the bottoms of the pools, state that they never find any tritons in the water during the winter months; but they discover great numbers of them in holes in the clay, and sometimes ten or

twelve coiled together. I have observed that either a very wet
or very dry situation is fatal to the triton during its state of
hybernation, and that a moderately damp one is always chosen
for that state of existence.

When in this state respiration is very low, and is
believed to be carried on through the pores of the
skin. No food is taken or required during this
period, and the body is comparatively stiff. Tritons
of the third and subsequent years usually hybernate
in company, a number of them being rolled together
into a lump as large as a cricket ball. Those of an
earlier period seem to descend deeper into the earth,
and hybernate singly.

When these batrachia are kept in confinement, it
will be supposed, from what has already been stated,
that the aquarium is *not* the best place for them
except in the tadpole state, or the breeding season for
mature specimens ; but that they should be kept in
a case under some arrangement that water may be
sought of their own will when desired, and that
during the winter some provision should be made for
their hybernation, under somewhat similar conditions
to those enjoyed in the natural state. To force them
at some periods of their life to exist in water is
something like endeavouring to compel eels to live
on land, and exchange their fishy habits for those of
a snake. The snake may love to lie in water sometimes, but it still remains terrestrial, and the eel

may be found on land, but is no less aquatic. The triton may be thoroughly aquatic at one period of its existence, and as completely terrestrial at another. Any attempt to subvert nature will only end in disappointment.

The amount of cold which the triton is able to bear is greater than one would suppose. On this point Mr. Higginbottom writes :—

I put two tritons into some water, and exposed them to a freezing temperature during the night ; in the morning I found the water frozen very firmly, with the tritons enclosed in its centre. On thawing they were lively and flexible. In the second experiment there was a piece of ice at the bottom of a circular vessel. I placed two tritons upon it, and then another covering of ice, and filled the vessel with water. I exposed it during the night in the open air to a temperature of 28° F. In the morning the whole had become a solid mass of ice twelve inches in circumference, with the animals in the centre. On breaking the ice carefully they were found completely encased in the ice. I had some difficulty in separating the extremity of one, but being liberated it used its arms and legs equally well.*

The Warty Newt is characterised by the following features. The skin is warted, and uniformly covered with scattered pores ; a row of pores occur on each side of the head, and along each side of the body, so as to form a line along the space between the fore and hind legs. A collar formed by a loose fold of the skin passes beneath the neck. The tail is very much flattened laterally, with sharp edges above and

* " Ann. and Mag. of Nat. Hist.," 1853, p. 378.

below, and terminates gradually in a blunt point. In spring a high membranous crest with a jagged edge runs along the back from between the eyes nearly to the tail, which also is crested. The upper parts are blackish-brown, blotched with rounded spots of a darker tint. The breast and belly are of a bright orange, or orange-yellow, with conspicuous round black spots, sometimes confluent, or running one into the other, and forming irregular bands. The sides are dotted with white, and there is often a silvery white band along the sides of the tail.

Entire length about five or six inches.

PLATE IX

Smooth Newt m & f

COMMON SMOOTH NEWT OR EFT.

MALE SMOOTH NEWT.

(*Lophinus punctatus.*)

THE Smooth Newt (Plate 9), Eft, or Effet, is the most common species of newt in the British Isles. It is found over a large part of Europe; is very common in Switzerland, Belgium, Galacia, and the Bukovina; uncommonly numerous in Silesia, and very abundant in Italy, especially near Rome; inhabits many parts of France, and is found in Carniola.* In our country it is found in almost every clear ditch or pond, sometimes in great numbers, where it feeds upon aquatic insects and

* Lord Clermont's " Quadrupeds and Reptiles of Europe."

worms, and in their turn, as some have observed, themselves become food for the great warty newt.

Though perfectly harmless, these poor creatures, devoid of any means of defence or offence, are often the subject of persecution in rural districts. School-boys, especially, consider them fair game for torture, and adults view its infliction with complacency, or at least without protestation, because not a few still entertain superstitious or fabulous notions either of their poisonous properties, or their secret association with the "black art." In towns, since aquaria, vivaria, and such "parlour menageries," have be-come fashionable, newts and lizards have been better known and appreciated, and have even been taken under the protection of the fair sex. In common with the warty newt, this species is slow in arriving at maturity, and there is every reason to believe that the remarks made under that species as to development, &c., will also apply to the present. In the first edition of Bell's "British Reptiles" considerable confusion was made of the smooth newts, on account of the different appearances pre-sented by them at different periods of their career. In the second edition most of the errors were cor-rected, but one was still maintained, to which Dr. Gray has referred in the following terms :—"Mr. Bell, believing that the form of the upper lip afforded a good character for the distinction of the species of these animals, divides them into two species, thus,—(1.) '*Lissotriton punctatus*,—upper lip straight, not overhanging the lower. (2.) *Lissotriton palmipes*,—upper lip pendulous at the sides, overhanging the under in a distinct festoon as far as the base of the lower jaw. Toes of the hinder feet fringed with a short membrane at all seasons.' I may observe that the latter is not the *Triton palmipes* of Latreille, which has the hind feet of the male, in the breeding season, webbed; and that I believe it only differs from the former by being in the fully developed state at the season of reproduction. I am borne out in this idea by the observations of Messrs. Higginbottom, Hogg, and many others. The former observes :—' Some tritons have been distinguished by the upper lip overhanging the lower.' I have observed that in the first year of *Triton asper* (*T. cristatus*) the upper lip overhangs the under," &c.*

Newts undergo confinement with complacency, and under such circumstances their insectivorous habits may render them useful. The results of one of these experiments are thus recorded :—

In the fern-case I formed a small pond of water, thinking that as effets are mostly found in ponds during the day, in

* Dr. Gray, in "Proceedings of the Zoological Society," 1858, p. 137.

summer they would enjoy the luxury of a bath. Not so, however. I never saw them voluntarily go into the water, and
when thrown in they always scrambled out as soon as possible.
The same thing occurs in keeping them in the aquarium ; they
always crawl out if they have the opportunity, seeming to
eschew the very element they are generally found in when
caught. These effets readily took food from the hand, particularly if it was rubbed against their noses ; they seemed almost
too sluggish to take much trouble in the matter else. I gave
them small worms, gentles, ants' eggs; they seized them with a
bite, and got them down with a series of gulps. But I hardly
ever fed them, perhaps not more than a dozen times altogether,
as my object was to combine business with pleasure ; the fern-
case, in common with most others I expect, being at times much
blighted with green fly. I first put in lizards, to try and keep
them down, but could not keep the lizards alive for any length
of time, owing, I think, to the dampness of the case not being
suitable to their constitutions, and their active habits making
them require more food than they could obtain. I then tried
small toads, and had the same luck with them as with the
lizards. My third venture, the newts, were a great success ;
they soon cleared off all the green fly within reach, crawling to
nearly the top of the ponds for that purpose. I have never had
any trouble with the green fly since their introduction to the
case.*

About the month of September, the smooth newts
seek a comfortable spot in which to pass their long
winter sleep. It may be under stones, bricks, or
pieces of timber, but is not often far beneath the
surface of the soil. On this occasion they may be
found, like the warty newt, in companies, closely
packed together :—

---

* *Science Gossip*, vol. i., p. 39.

Rolled up like a ball,
In a hole snug and small,
They sleep till warm weather comes back, poor things.

Though the children in the nursery sing thus of the dormouse, it is equally applicable to the newts. In the early spring they emerge from their places of concealment, and the mature animals seek the water to pass a tripled honeymoon in an aquatic state.

The eft, in common with other reptiles, casts its skin at certain, or perhaps uncertain, periods of the year. Mr. Guyon has thus described the process, from his own observation :—

The operation was nearly completed, the skin being pushed down the body in a ring, by which the hinder legs were, to use an Irishism, handcuffed to the tail. The snout was principally used in pushing it down, and the tail was scarcely free when the animal seized the skin with its mouth, and in half-a-dozen gulps swallowed it. This act occupied nearly a minute, during which three filmy gloves, the integuments of the paws, were projecting from the mouth. Although a tremendous yawn testified to the fatigue of the performance, the newt made no objection to concluding the meal with a scrap of roast mutton.*

From Mr. Higginbottom's remarks one might conclude that the mode of depositing the ova was the same in this instance as in that of the warty newt, and that each ovum was deposited separately in the fold of a leaf. From this conclusion Mr.

* *The Zoologist*, p. 6210.

Kinahan dissents, and observes that he has seen the
ova of the smooth newt deposited in strings of from
four to six about the roots of aquatic plants, and
that such ova he has hatched. Our own observations confirm those of Mr. Kinahan, as we have seen
them deposited in similar strings or chains, on
stones, &c., but have never attempted to preserve or
place them in conditions for hatching. Three of
the species of newt herein recorded are said to be
inhabitants of Ireland, and of these the present
species is most common. The warty newt is more
doubtful, and the palmated newt was found by Mr.
Thompson in the western wilds. There is no doubt
that all are condemned to death as soon as they
meet the eye of a thorough-going Irishman.

In a paper by Mr. Kinahan, read before the
Dublin Natural History Society (Feb. 10, 1854),
the superstitions regarding these reptiles hold a
conspicuous place. " In almost every part of the
country we find these animals," he says, " looked on
with disgust and horror, if not with dread. This
arises from two superstitions : one of them, common
to great part of Ireland, relating chiefly to the
animal in its aquatic state, and which in the county
of Dublin has earned for it the names of Man-eater
and Man-keeper; though the 'dry ask' of the
county of Dublin—that is, the animal in its terrestrial stage—is supposed to be equally guilty with

the first-mentioned in the habit of going down the
throats of those people who are so silly as either to
go to sleep in the fields with their mouths open, or
to drink from the streams in which the dark
'lewkers' (newts in the aquatic state) harbour: they
are also said to be swallowed by the thirsty cattle.
In consequence, the country people kill them whenever they meet with them on land, and poison the
stream they are found in, by putting lime into the
cattle's drinking-pools. In either case the result is
the same, the reptile taking up his quarters in the
interior of his victim in some way, it would puzzle a
physiologist to explain how. It contrives to live on
the nutriment taken by the luckless individual or
animal, so that, deprived of its nourishment, the
latter pines away; nay, so comfortable does the
newt make herself that, not content with living by
herself, she contrives to bring up a little family.
Often have I been told of the man who got rid of a
mamma newt and six young ones by the following
recipe, which I am assured is infallible:—The
patient must abstain from all fluids for four-and-
twenty hours, and eat only salt meats; at the
expiration of that time, being very thirsty, he must
go and lie open-mouthed over a running stream—
the noisier the better; when the newts, dying of
thirst, and hearing the music of the water, cannot
resist the temptation, but come forth to drink, and

of course you take care that they do not get back
again. The dry ask, in addition to this bad
character, is also supposed to be endowed with the
power of the ' evil eye '—children and cows exposed
to its gaze wasting away. The Rev. J. Graves states
that in Kilkenny it is looked on as 'a devil's beast,'
and, as such, burnt. But to compensate in some
measure for its evil qualities, the dry ask is said in
Dublin to bear in it a charm. Any one desirous of the
power of curing scalds or burns has only to apply
the tongue along the dry ask's belly to obtain the
power of curing these ailments by a touch of that
organ. In the Queen's County it is also used to
cure disease, but in a different way; being put into
an iron pot under the patient's bed, it is said to
effect a certain cure, though of what disease I am
not quite clear."*

Lord Clermont's excellent description of this
reptile may usefully close this chapter :—

The whole of the skin is quite smooth, without
any tubercles ; on the top of the head are two rows
of pores ; occasionally there are a few distinct pores
on the sides, forming an indistinct lateral line ; the
collar beneath the throat very inconspicuous; the
male in the breeding season furnished with a crest,
which runs continuously from the top of the head

* _The Zoologist_, p. 4355.

along the tail, and is regularly festooned on its edge. The upper parts are of a light brownish-grey, inclining to olive; yellowish beneath, becoming bright orange in spring, marked all over with round, black, unequal spots; on the head the spots form about five longitudinal streaks; and there is a yellowish streak under the eyes. The female is much less spotted than the male, the spots being smaller and often very obscure, and the under parts are often quite plain. It passes a great deal of its time on land, when the skin loses its softness and sometimes becomes wrinkled; the toes, from being flat, become round; the membranes of the back and tail entirely disappear, and all the colours become more dull.*

Entire length, from 3½ to 4 inches.

* Lord Clermont's "Quadrupeds and Reptiles of Europe." p. 264.

FEMALE SMOOTH NEWT.

PALMATE NEWT.

(*Lophinus palmatus*, Dum. and Bibr.)

THIS species (Plate 10, Fig. 3) is said to be the most common newt around Paris, and Lord Clermont states that it occurs in many parts of Germany, near Vienna, and elsewhere. It is common in the south of France, but rare in Switzerland. In Italy it has been found near Rome and about Pisa. To this M. Deby adds more explicitly that it was found " by Fournelle in the department of the Moselle, by Sturm in Germany, by Razoumowski in Switzerland, by Latreille in France, and by De Selys, Van Haesendonck, and myself, in Belgium." *

The specimens described and figured in the first edition of Bell's " Reptiles " as the Palmated Smooth

* *The Zoologist*, p. 2231.

PLATE V

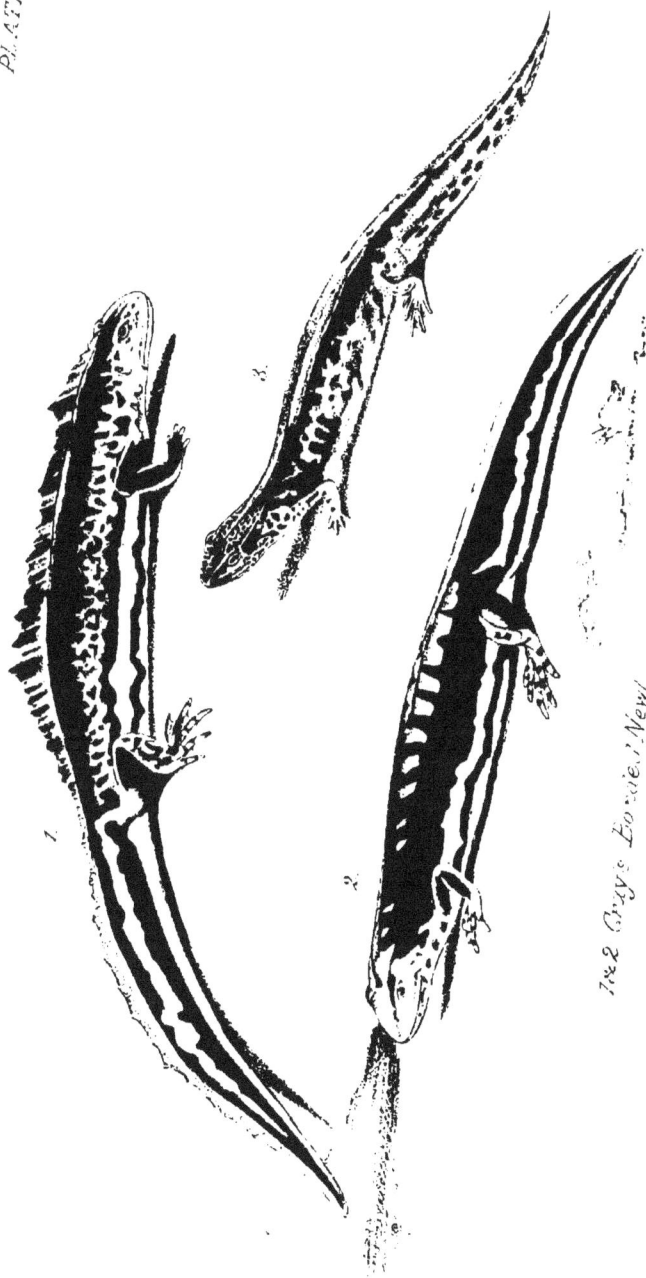

1 & 2 Gray's Border Newt

Newt (*Lissotriton palmipes*), do not represent a genuine species at all, but a variety of the common newt, spotted on the sides and belly with large black spots. This error was corrected in the second edition, and new figures, not very characteristic, added.

This species appears to have been first recognized by Razoumowski in 1788, when it was discovered in the fountain of Vernens, in the canton of Vaud, in company with the Smooth Newt. The first notice of it in this country, as far as we are aware, is the account in the *Zoologist*, dated May 3rd, 1848, by Mr. J. Wolley, in which he mentions having found it around Edinburgh ; and that during a ramble in the Pentland Hills he saw no other species.* In the succeeding number of the same journal, Mr. W. Baker, of Bridgewater, stated that he had found this species in the neighbourhood of Bridgewater for many years, and had forwarded specimens to Mr. Bell ; † so that the merit of its discovery rests with Mr. Baker. A subsequent communication from Mr. Wolley adds another locality, that of some pools by the side of the hills which rise from Loch Eribol, on the west.‡

About this time M. Julian Deby indicated the points of difference between the common newt and the present species, with a view to the correction of

---

* *The Zoologist*, p. 2149.   † *The Zoologist*, p. 2198.
‡ *The Zoologist*, p. 2265.

M

those who had maintained that the Palmate Newt
was only a variety of the common newt; whilst Mr.
Bell had announced his belief in its being a species
entirely new to science. We believe that Mr.
Edward Newman first declared that it was specifi-
cally distinct from *Lophinus punctatus*, and, moreover, that it was not "new to science," but was
really the true Palmated Newt.

Five-and-twenty years ago, Mr. Holdsworth says,
in company with other small boys, he used to catch
black-footed newts in a pond near Dartmouth, in
Devonshire; the means of capture being of the
simplest kind, consisting of a bit of twine fastened
to a small bent pin, and a worm for bait. He had
since caught a great many of these newts, and three
years ago (1860) they were abundant in the same
pond. In 1863 he spent a few days in Hereford-
shire, and near the village of Letton, about twelve
miles from Hereford, had again the satisfaction of
seeing *Lissotriton palmipes* in abundance, although,
as far as he could ascertain, confined to one pond.
In this case, as well as at Dartmouth, *L. palmipes*
was the only species to be found. A number of
specimens were sent by him to the Zoological Gardens, and at a meeting of the Zoological Society
specimens were exhibited.* Its habits do not ap-

---

* See "Proceedings of Zool. Soc., 1863," p. 159

parently differ from those of the common smooth
newt. The' males show the same lateral curvature
of the tail, with a rapid vibration of its lash-like
extremity during the love season. The slough is
cast entire, and, in most cases, immediately swallowed by its owner.*

In pointing out the distinctions to be observed
between this species and its allies, Professor Bell
notes that "the whole animal is smaller; the head
flatter, broader in proportion, and beautifully mar-
bled. The crest is straight, and much less elevated
than in the other species, and begins further back
on the neck. The hinder feet of the male are palmate; entirely so in the summer, less so in the
autumn, and towards winter the web is scarcely
broader than in the smooth newt in the full season.
The tail is not much more than half the depth, terminating rather abruptly, and furnished at its
extremity with a small filament, which varies in
length from two to four lines, in the female dwind-
ling to a mere mucronation. The colours of the
back and sides are more clear and bright, although
generally darker. The spots are more numerous
and often confluent; and the tail has two distinct
longitudinal fasciæ of spots, with occasionally a few
between them; but the inferior margin is invariably

* E. W. H. Holdsworth, *Zoologist*, p. 8640.

M 2

and distinctly pale and immediate. The female is usually paler than the male; but the spots on the tail are in general more numerous, smaller, and disposed to become confluent."* Indeed, so distinct and permanent are the characters which separate the present species from its congeners, that it seems surprising that it remained so long without recognition.

In a communication made by Dr. Gray to the Zoological Society in 1863, he thus adverts to some points of difference between this species and others :—

The *T. cristatus* has a circular ring-like iris, and the only Batrachians which appear to have the spot on each side of the iris, forming a band across the eyes, are the English *Lophinus punctatus* and *L. palmatus;* the band on the eyes looking in these like a continuation of the dark streak on the side of the head. I may add that the best character for the distinction of these two species, which are often found in the same pond, is that in *L. punctatus* the crest of the male is scalloped on the edge, and high in front, while in *L. palmatus* it is low in front and higher behind, and has a smooth straight upper edge. The tail of the latter is also always truncated, and usually appendaged at the tip.†

M. Deby thus compares the two species :—

L. punctatus	*L. palmatus.*
1. Tail generally tapering to a point.	1. Tail suddenly truncate before the apex, and terminating in a slender filament three lines in lenth.

* Bell's " Brit. Rept.," 2nd ed., p. 156.
† Dr. Gray, in " Proc. Zool. Soc., 1863," p. 203.

2. Hind feet having the toes free, only edged by a membrane.	2. Hind feet perfectly palmate, all the toes united by a membrane.
3. Back with a very large festooned undulating crest, which extends from the nape of the head to the end of the tail. No lateral elevated ridges.	3. Back flattened, with two elevated lateral lines passing above the eyes, and extending to the base of the tail. The dorsal crest small and simple.
4. Length much greater than *palmatus*.	4. Size much smaller than *punctatus*.*

The females are more difficult to distinguish from the same sex of the common smooth newt than are the males; but even these present features which are characteristic, and which were indicated by Mr. Wolley as supplementary to the above. Their heads seem broader and shorter than in *L. punctatus*, and the toes of their hind feet are, for the most part, shorter, the males also have the former but not the latter character. As to the colour, if in a genial situation, the belly is usually a delicate milk-and-water white, tinged more or less with yellow towards the middle line; the back and sides of the body and tail are of a dark olive-green, and in some, particularly very large specimens, are beautifully mottled by a network of lighter colour. In moor land the skin becomes harsh, and coloured more like the females of the common newt, sometimes even to the orange belly.†

* *The Zoologist*, p. 2232.
† *The Zoologist*, p. 2268.

The distinguishing features of this species are, besides belonging to the smoothed-skinned section, that there is a prominent line running down on each side of the back from the nose to the hind legs. The crest, or fin, along the back and tail is highest on the tail. The toes of the male are united and completely webbed. The upper part of the body of the male is olive-brown or greenish, with dark spots, with a wide band of yellowish white, bordered with round spots, on the side of the tail; and the belly is yellow, with a few darker spots. The female is lighter coloured, and differs so much in general appearance in the spring that it has been described as a distinct species.

Mr. Higginbottom thus summarizes the characteristics of this species :—

1st. Tail suddenly truncate before the apex, and terminating in a slender filament three lines in length.

2nd. Hind feet perfectly palmate, all the toes united by a membrane.

3rd. The dorsal crest small and simple.

4th. Size much smaller than the smooth newt (*Lophinus punctatus*).

I have fully ascertained the changes when the breeding season is over. The slender filament is absorbed, and the truncated portion of tail becomes obtusely rounded off, with a slight indurated dark tip at the end, and the web on the hind feet is wholly absorbed, leaving the toes free.*

* "Annals of Natural Hist.," 2nd series, vol. xii., p. 382.

TADPOLE OF NEWT.

GRAY'S BANDED NEWT.

(Ommatotriton vittatus, Gray.)

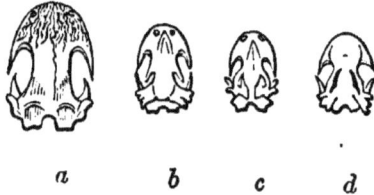

a   b   c   d

SKULLS OF NEWTS.*

THE fourth species of British Newt (Plate 10, Fig. 1 and 2) rests alone on the authority of specimens in the British Museum, and is now considered more than doubtful, but, with this reservation, we have continued to include the record. The sole habitat of this species for some time known was Lycia, with the exception of the specimens in the British Museum, said to have been collected by Dr. Gray in the neighbourhood of London; but more recently it has been found to inhabit Holland, Belgium, and France. Professor Bell described and

* *a.* Triton cristatus.   *b.* Lophinus punctatus.
   c. Lophinus palmatus.   *d.* Ommatotriton vittatus.

figured one in his "British Reptiles" as a variety
of his Palmated Smooth Newt, which latter now
proves to be a variety of the Common Smooth
Newt.

The present newt differs generically from the
Common Newt, that is, the difference is of such a
nature as to warrant its exclusion from the same
genus which includes the Common Newt, and its
being placed in another, and a new genus, under
the name of *Ommatotriton*. From his remarks, it
would appear that Professor Bell considered the
specimens above alluded to as veritably indigenous,
for he says, "There is no reason to doubt that they
are British, and there is ground for believing that
they were taken at no great distance from London."
His sole objection seemed to be that he did not
regard them as different from his own Palm-footed
Newt, and he quotes Dr. Gray's letter to him on the
subject, which would probably not be so decided
then as now; for, as it will be hereafter seen, Dr.
Gray was more convinced than ever that he was
right, and that Professor Bell was wrong.

The letter quoted is to the following effect:—"My
Salamandra vittata (meaning the present species),
which has been figured by Guérin,* who has adopted
my name, belongs to the same group as the former

* Guérin, "Icon. Règ. Animal," 17. t. 28. f. 2.

(*Triton cristatus*). It agrees with it in having the crest interrupted over the loins, and chiefly differs from it in having smaller tubercles, and in colour. It is easily known both from *S. palustris* (*Triton cristatus*) and from *Triton* (*Lophinus*) *punctatus* by the wide black-edged white streak along the lower part of each side of the body, &c. The head is much larger, and more depressed, than that of any of the varieties of *T. punctatus*.

" The species is found in Holland and Belgium as well as here. It must be very local in this country, as I have seen no specimens since those I caught some thirty years ago."

Whilst debating as to the propriety of including this species amongst " Our Reptiles," we communicated with Dr. Gray on the subject, to which he replied :—" *Triton vittatus*, Gray, is not only a distinct species, but a distinct genus from any of our European *Tritons*, characterized by the form of the skull and the large size of the orbits."

On referring to a paper on the Salamandrines, published by Dr Gray in 1858, we found that after giving England, the North of France, and Belgium, as localities for this species, he makes the following observations :—" Mr. Bell, in his ' British Reptiles,' gives a good figure of one of my specimens of this species, which he is convinced ' is to be considered as a variety of the Palmate Newt. The osteological character, as well as the form of the dorsal crest, and the disposition of the colours, shows this is not the case, and that it is not only a distinct species, but a very distinct genus, as is further proved by M. Dugès' figure of the skull.'"* In illustration of the latter remark, we have given wood-cuts of the skull of this species, together with the other British Newts, which have been faithfully copied from M. Dugès' treatise in the French "Annals of Natural History."†

The habits of this species do not probably differ from those of the Smooth and Warted Newts; but hitherto it has been found so seldom, and the observations have been so few, that beyond its merely scientific character, we have no character to give it. Indeed, in a popular sense we may almost say that it is a newt without a character. So far as genus and species are concerned, it is regarded as very well characterized by scientific men. In this sense we will give its character, as written by Dr. Gray.

"Pale grey, closely black-spotted. Tail nearly black. Side of abdomen and middle of tail with a broad wide streak, white beneath (belly). Throat black-dotted. Mature male, during the breeding season, with a high-toothed dorsal (back) and caudal

---

* Dr. Gray, in " Proc. Zool. Soc., 1858," p. 140.
† *Ann. des Sc. Nat.*, 3rd series, xvii. t. 1.

(tail) fin, the base interrupted over the loins. Toes separate, webbed, slightly finned."

Our figure is taken from one of Dr. Gray's specimens in the British Museum, courteously placed at our disposal for this purpose.

Here terminate the Amphibia, as represented in the British Islands. It has already been observed that most excellent herpetologists consider the Amphibia in the light of a *class* distinct from the Reptilia. We have no doubt that this distinction will continue to be maintained, as it appears that recent investigations and discoveries tend more and more to the dissociation of these two groups of reptiles.

TADPOLE OF NEWT.

PLATE XI

1 Hawks Bill Turtle.

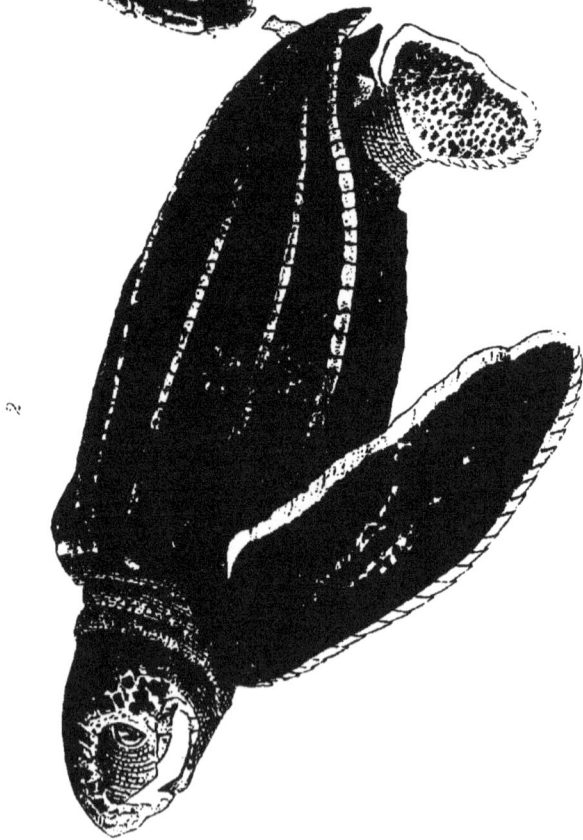

2 Leathery Turtle.

THE HAWK'S-BILL TURTLE.

(*Chelonia imbricata*, Schw.)

WE have advisedly reserved the Chelonians until
the close of "Our Reptiles," instead of commencing
with them, as in true scientific arrangement we
were almost bound to do. It is a doubtful point
with us whether they deserve a place at all, certainly
not a prominent one; and that this little volume
might not be regarded as incomplete in its enumer-
ation, we have added them at the end, in the hope
of " saving our turtle," even though it should prove
to be only " mock turtle."

The Hawk's-Bill Turtle (Plate 11, Fig. 1) is an
American species, and its occurrence on our coasts
must be regarded as the result of accident. Sibbald
refers to one, of which he possessed the shell, and
which came from the Orkneys. Dr. Fleming
states :—" I have credible testimony of its having
been taken at Papa Stour, one of the West Zetland
Islands," and Dr. Turton records an instance of one
which in the year 1774 was taken in the Severn,

and placed in his father's fish-ponds, where it lived
till winter. In 1849 a large specimen was found
floating on the sea, off Redcar, but it was dead.
This is all the evidence, slender as it may be, upon
which the present species is included in the Fauna
of Great Britain. This we shall not in the present
instance much regret, as it is a very interesting
species, and one which possesses a commercial value,
so that we shall avail ourselves of the example of
our predecessors, and include it amongst our
number.

The Chelonians are divided into families for the
purposes of classification; one of these includes the
Thalassians, or Sea-Turtles, to which both species
we shall have occasion to enumerate belong.
In this family the fore-limbs are considerably
lengthened, and all are modified into flappers, or
paddles, for swimming. Turtles seldom leave the
sea, except to deposit their eggs, though some
accounts state, "they will crawl up the shores of
desert islands in the night, and clamber up the
edges of isolated rocks far at sea, for the purpose of
browsing on certain marine plants." They may be
met with far out at sea, floating motionless upon the
water, as if dead. They are good divers, and can
remain a long time beneath the surface of the
water. Their food consists chiefly of marine plants,
whilst some of them do not object to a delicate

crustacean, or to appropriate molluscs as their chief articles of food.

At the breeding season these reptiles seek a retired, sandy beach after sunset, and there, above high-water mark, a hollow is excavated to serve as a nest, in which during the night about a hundred eggs are deposited. It is said that the female repeats this operation three times in the same season, at intervals of two or three weeks. The eggs are lightly covered with sand, and left to be hatched by the heat of the sun, which operation is accomplished in two or three weeks. The parents thenceforth take no regard of their progeny. As soon as the young turtles leave the egg, they seek the sea, some of them falling a victim to rapacious birds during the brief journey they have to perform, and many others serving as a delicate morsel for predatory fish as soon as they reach the water. Thus the balance of life is maintained, or the surface of the sea *might* soon be covered with floating turtles.

The eggs of the turtles are more or less spherical, and are much prized as articles of food. A native Brazilian will eat twenty or thirty for breakfast, as they are about the size of those of a " Bantam " hen. By the river Amazon a large number of turtle eggs are secured every year for the sake of turtle-oil. The stratum of eggs in the sand is ascertained by a pole thrust in, and the harvest of eggs is

estimated, like the produce of a well-cultivated acre ;
an acre accurately measured, of 120 feet long and
30 wide, having been known to yield one hundred
jars of oil. The eggs when collected are thrown
into long troughs of water, and being broken and
stirred with shovels, they remain exposed to the sun
till the yolk, the oily part, is collected on the surface, and has time to inspissate ; as fast as this oily
part is collected on the surface of the water, it is
taken off and boiled over a quick fire. This animal
oil, or " tortoise-grease," when prepared, is limpid,
inodorous, and scarcely yellow, and it is used, not
merely to burn in lamps, but in dressing victuals, to
which it imparts no disagreeable taste. It is not
easy, however, to produce oil of turtle's eggs quite
pure ; there is generally a putrid smell, owing to
the mixture of addled eggs. The total gathering of
the three shores between the junction of the Orinoco
with the Apure, where the collection of eggs is
annually made, is 5,000 jars, and it takes about
5,000 eggs to furnish one jar of oil.* By this one
means, therefore, in a single district, twenty-five
millions of turtles are prevented coming into existence every year ; or as many as would, when full
grown, cover eight square miles, as closely as they
could be placed to each other.

* Simmonds' "Curiosities of Food," p. 182.

Turtle-catching is another means by which excessive turtle-population is kept in check—whether
to supply the tables of epicures with the dainty
" green turtle," or to secure the horny plates of the
" Hawk's Bill " for conversion into combs, card-cases,
&c., for the use of the *élite* of the civilized world.
One method adopted is to watch the females as they
come on shore to deposit their eggs, to turn them
on their backs, and let them lie, helplessly vibrating
their " flappers," till their hunters think fit to kill
them, or carry them away. Another method is
described by Mr. Darwin as adopted at Keeling
Island. The water is so clear and shallow that,
although at first a turtle quickly dives out of sight,
yet in a canoe or boat under sail the pursuers, after
no very long chase, come up to it. A man standing
ready in the bows at this moment dashes through the
water upon the turtle's back ; then clinging with
both hands by the shell of the neck he is carried
away till the animal becomes exhausted, and is
secured.*   A more curious mode of capture is that
by means of the *Remora,* or sucking-fish. A number of these fish are carried in tubs in the boats
that go in search of turtles. To the tail of each
fish is attached a strong cord. When the fishermen
perceive the turtle basking on the surface of the

---

* Darwin's " Journal of Researches."

N

sea, they drop one of these fish overboard, retaining
hold of the cord, which they "pay out" to a sufficient length not to impede the *Remora*. As soon
as the fish observes the floating reptile it makes
towards it, and by means of its sucker attaches
itself to it so firmly that both fish and turtle are
drawn into the boat. Several other methods are
adopted in other localities, but these are amongst
the most curious.

The tortoise-shell of commerce is in part yielded
by this species, although there is little doubt that
both in ancient times and in the present, more than
one, and probably several species, afforded tortoise-
shell.

Bruce, the African traveller, alludes to this
article —

The Egyptians (he says) dealt very largely with the Romans
in this elegant article of commerce. Pliny tells us, the cutting
them for veneering or inlaying was first practised by Carvilius
Pollio, from which we should presume that the Romans were
ignorant of the art of separating the laminæ by fire placed in the
inside of the shell, when the meat is taken out; for these
scales, though they appear perfectly distinct and separate, do
yet adhere, and oftener break than split, where the mark of
separation may be seen distinctly. Martial says that beds were
inlaid with it. Juvenal and Apuleius, in his tenth book, mentions that the Indian bed was all over shining with tortoise-shell
on the out-ide, and swelling with stuffing of down within. The
immense use of it in Rome may be guessed at by what we learn
from Velleius Paterculus, who says that when Alexandria was
taken by Julius Cæsar, the magazines or warehouses were so full
of this article that he proposed to have made it the principal

ornament of his triumph, as he did ivory afterwards, when
triumphing for having happily finished the African war. This, too,
in more modern times, was a great article in the trade to China,
and I have always been exceedingly surprised, since near the
whole of the Arabian Gulf is comprehended in the charter of the
East-India Company, that they do not make an experiment of
fishing both pearls and tortoises; the former of which, so long
abandoned, must now be in great plenty and excellence; and a
few fishers put on board each ship trading to Jidda, might surely
find very lucrative employment with a long boat or pinnace, at
the time the vessels were selling their cargo in the port; and
while busied in this gainful occupation, the coasts of the Red
Sea might be fully explored.

The thirteen plates which cover the carapace are
known technically as two plates, two main plates,
three backs, two wings, two tongues, and two shoul-
ders—in all, thirteen. In an animal of the ordinary
size, about three feet long and two and a half wide,
the largest plates will weigh about nine ounces, and
measure about twelve inches by seven, and one-
fourth of an inch thick in the middle.*

The market value of tortoise-shell is subject to
many fluctuations, but at ordinary times it ranges
from sixteen to thirty shillings per pound. The
total value of our imports in one year is estimated at
about twenty-five thousand pounds. The greater
proportion of this is derived from the British East
Indies, and the residue from Central and South

* *The Technologist*, vol. i., p. 382.

America, Australia, Egypt, the West India and
Philippine Islands.

The mode of working this substance is described
at some length by MM. Dumeril and Bibron, in
their large work on " Reptiles." In order to straigh-
ten the plates, which, when detached, are bent in
various ways, it is sufficient to steep them in boiling
water for a few minutes, and then take them out
and place them between plates of metal or smooth
blocks of hard wood, leaving them to cool, great
pressure being employed at the same time. They
then retain the flatness desired. They are next
scraped and filed, a smooth surface being obtained
with as little loss as possible. When these shells or
scales are brought to a proper thickness and size,
they may be then used separately; but they are
generally submitted to a still further preparation.
When, for instance, they are too thin, or when they
are not sufficiently long or broad, the following processes are employed:—In order to obtain single
plates of great size, two are soldered together, the
thin part of one being laid upon the thin part of
the other; or, as is sometimes done, the edges of
each plate are delicately bevelled and fitted together.
In each case they are then put between metallic
plates; to these a certain degree of pressure is
given, which, when the whole is plunged into boiling
water, is increased, and by this mode they are so

intimately joined together that the slightest trace
of their union cannot be detected.

It is almost exclusively by means of boiling water
that the effects upon tortoise-shell are obtained.
The substance of the scales becomes so softened by
the action of the heat, that it may be acted upon
like a soft mass, or a flexible and ductile paste,
which, by pressure in metallic moulds, will assume
every variety of form required. The soldering of
two pieces together is effected by means of hot
pincers, which, while they compress, at the same
time soften the opposed edges of each piece, and
amalgamate them into one. No portion of the
scales is worthless; the raspings and powder produced by the file, mixed with small fragments, are
put into moulds, and subjected to the action of
boiling water, and thus made into plates of the
desired thickness, or into various articles which
appear as if cut out of the solid block.

The following is Dr. Turton's description of the
present species:—

"Body roundish-ovate, slightly heart-shaped,
slightly carinated down the back; head small,
prominent, with the upper mandible curved over
the lower; two claws on each foot; plates of the
disk imbricated, thirteen in number, rather square,
semi-transparent, variegated; of the circumference
twenty-five, pointed, and incumbent on each

other in a serrated manner ; tail a mere notch." *

General length from the tip of the bill to the end of the shell, about three feet ; has been known to measure five feet.

* Turton's " Brit. Fauna," p. 78.

THE LEATHERY TURTLE.

(Sphargis coriacea.)

THIS Turtle (Plate 11, Fig. 1) differs materially
from the preceding in several points. The carapace,
instead of being clad with plates, is covered with a
tough, leather-like skin. Along the upper shell are
seven distinct ridges, running longitudinally; these
are sharp and slightly toothed in the adult animal,
but rounded in the young. It appears to be a
native of the Mediterranean Sea, but even there is
not at all common; it is found also in the Pacific,
Atlantic, and Indian Oceans. The claims of this
turtle to be regarded as British are founded upon a
passage in Borlase's " History of Cornwall," and
Pennant's " British Zoology." Borlase records that
two were taken on the coast of Cornwall in the
mackerel nets, of a vast size, a little after Midsum-
mer, 1756; the largest weighed eight hundred
pounds, the lesser near seven hundred. Pennant
adds, that a third, of equal weight with the first of
the above, was caught on the coast of Dorsetshire,

and deposited in the Leverian Museum. The largest of the Cornish specimens measured six feet nine inches from the tip of the nose to the end of the shell, and ten feet four inches from the extremities of the fore fins extended. A specimen, eight feet long, was caught in a herring net, at Bridlington Quay on the 15th October, 1871.

Shaw mentions a specimen taken on the French coast, in the month of August, 1729, about three leagues from Nantz, not far from the mouth of the Loire, and which measured seven feet one inch in length, three feet seven inches in breadth, and two feet in thickness. It is said to have uttered a hideous noise when taken, so that it might be heard to the distance of a quarter of a league ; its mouth at the same time foaming with rage, and exhaling a noisome vapour."* The bellowing noise made by the members of this genus led to the adoption of the generic name, which .is derived from the Greek σφαραγέω, " to make a noise in the throat."

The turtle so essential to the comfort of an alderman is not this species. More than one kind is regarded as very good eating, but the true Green Turtle is *Chelonia Mydas*. It is said that the Leathery Turtle is positively injurious. Pennant narrates an instance in his " Appendix to British

---

* Shaw's " General Zoology," vol. iv., part i., p. 78.

Zoology ":—"The late Bishop of Carlisle informed
me that a tortoise was taken off the coast of Scar-
borough in 1748 or 1749. It was purchased by a
family at that time there, and a good deal of company invited to partake of it. A gentleman who
was one of the guests told them it was a Mediterranean turtle, and not wholesome; only one of the
company ate of it, and it almost killed him, being
seized with a dreadful vomiting and purging."

The introduction of the Green Turtle to this
country as an article of food is of comparatively
recent date, probably not much exceeding a century,
and it is still confined to a limited circle of ad-
mirers. The plebeian eye may gaze with longing
on "turbot," but does not so often flash with desire
for turtle soup. The green fat is an aristocratic
delicacy, a taste for which is acquired best by means
of an aldermanic gown.

Of the Sea Turtles, the most in request, says
Catesby, is the Green Turtle, which is esteemed a
most wholesome and delicious food. It receives its
name from the fat, which is of a green colour. Sir
Hans Sloane informs us, in his "History of Jamaica,"
that forty sloops are employed by the inhabitants of Port Royal, in Jamaica, for catching them.
The markets are there supplied with turtle as ours
are with butcher's meat. The Bahamians carry
many of them to Carolina, where they turn them

to good account, not because that plentiful country
wants provisions, but they are esteemed there as a
rarity, and for the delicacy of their flesh. They
feed on a kind of grass, growing at the bottom of
the sea,* commonly called turtle-grass. The inhabitants of the Bahama Islands, by frequent practice,
are very expert at catching turtles, particularly the
Green Turtle. In April they go, in little boats, to
Cuba and other neighbouring islands, where, in the
evening, especially in moonlight nights, they watch
the going and returning of the Turtle to and from
their nests, at which time they turn them on their
backs, where they leave them, and proceed on turning all they meet ; for they cannot get on their feet
again when once turned. Some are so large that it
requires three men to turn one of them. The way
by which the Turtle are most commonly taken at
the Bahama Islands is by striking them with a
small iron peg of two inches long, put in a socket,
at the end of a staff of twelve feet long. Two men
usually set out for this work in a little light boat or
canoe, one to row and gently steer the boat, while
the other stands at the head of it with his striker.
The Turtle are sometimes discovered by their swimming with their head and back out of the water;
but they are oftenest discovered lying at the

---

* It is not a grass, but the sea-wrack (*Zostera marina*), which
is here alluded to.

bottom, a fathom or more deep. If a Turtle perceives he is discovered, he starts up to make his escape, the men in the boat pursuing him, endeavouring to keep sight of him, which they often lose, and recover again by the Turtle putting his nose out of the water to breathe; thus they pursue him, one paddling or rowing, while the other stands ready with his striker. It is sometimes half an hour before he is tired; then he sinks at once to the bottom, which gives them an opportunity of striking him, which is by piercing him with an iron peg which slips out of the socket, but is fastened with a string to a pole. If he is spent and tired by being long pursued, he tamely submits, when struck, to be taken into the boat or hauled ashore. There are men who, by diving, will get on their backs, and by pressing down their hind parts, and raising the fore parts of them by force, bring them to the top of the water, while another slips a noose about their necks.*

To return to the more immediate subject of this chapter, it is supposed that the Leathery Turtle was the species which supplied Mercury with its back shell, to which he applied strings, and thus extemporized the first *lyre*, which was the prototype of all stringed musical instruments. The seven dorsal

---

* Catesby's "Natural His'ory of Carolina."

ridges to which we have alluded strengthened this
supposition ; the ancient lyre having, according to
some writers, that number of strings.

The carapace is heart-shaped, the hinder extremity much pointed; an elevated ridge follows
the dorsal line from end to end ; and on either side
of this central ridge are three parallel ones, equi-
distant from each other; between these ridges the
surface is quite smooth ; the head is without plates;
the jaws are very strong ; the lower jaw turns
upwards at its extremity, forming a hook, which is
received into a corresponding channel in the upper
jaw.   In the young, the lines on the carapace are
formed by a succession of tubercles in rows, and the
entire surface, both of it and of the plastron, is
warty.   The eyelids are divided almost vertically ;
the fore feet, or fins, are as long again as the hinder,
the latter, however, being the wider ; there is no
trace of nails to the toes; the tail is as long as the
point at the hinder extremity of the carapace.   The
general colour is brown, with numerous pale yellow
spots on the upper surface ; the legs and tail are
black.

The entire length sometimes exceeds six feet.*

* Clermont's "Quadrupeds and Reptiles of Europe," p. 169.

APPENDIX.

CLASSIFICATION AND SYNONYMS OF BRITISH SPECIES.

Class.—REPTILIA.

Order I.—CHELONIANS.

Family 5.—Cheloniadæ (*Sea Tortoises*).

1. CHELONIA IMBRICATA. *Schw.* Hawk's-Bill Turtle. *P.* 173
   Caretta imbricata. *Gray.* Syn. Rept. i. 52.
   Testudo imbricata. *Linn.* Syst. Nat., p. 350
   Testudo Caretta. *Knorr.* Delic. ii. 124
   Eretmotchelys imbricata. *Fitz.* Syst. Rept. 30
   Chelonia imbricata. *Schw.* Prod. 291
   Chelonia multiscutata. *Kuhl.* Beytr 78
   Chelonia pseudo-caretta. *Less.* Voy. Bel. 302
   Scaled Tortoise. *Grew.* Mus. 38
   Imbricated Turtle. *Shaw.* Gen. Zool. iii., 89
   Hawk's-Bill Turtle. *Bell.* Brit. Rept., p. i.
2. SPHARGIS CORIACEA. *Gray.* Leathery Turtle. *P.* 183
   Sphargis Mercurialis. *Merr.* Amph. 19
   Sphargis Mercurii. *Risso.* Hist. Nat. 85
   Testudo coriacea. *Linn.* Syst. Nat. 350

Testudo Mercurii. *Gesn.* Aquat. 1134
Testudo lyra. *Bechs.*
Testudo tuberculata. *Penn.* Phil. T,ans. lxi.
Dermatochelys atlantica. *Leseur.* Cuv. R.A. ii., 14
Dermatochelys porcata. *Wagl.* Syst. Amph. 133
Coriudo coriacea. *Flem.* Brit. Anim. 149
Turtle. *Borlase.* Nat. Hist. Corn. 285
Coriaceous Tortoise. *Penn.* Br. Zool. iii., p. 7
Coriaceous Turtle. *Shaw.* Gen. Zool. iii., p. 77
Leathery Turtle. *Bell.* Brit. Rept. 12
Spinose Tortoise. *Penn.* Brit. Zool. iii.
Tuberculated Tortoise. *Penn.* Phil. Trans.
Trunk Turtle.

ORDER II.—SAURIANS.

FAMILY 1.—LACERTINIDÆ (*Lizards*).

3. ZOOTOCA VIVIPARA. Viviparous Lizard.   *P.* 23

Lacerta vivipara. *Dum. and Bib.* Rept. v., 204
Lacerta crocea. *Wolf.* Sturm Faun., t. 4
Lacerta nigra. *Wolf.* Fauna Germ.
Lacerta pyrrogaster. *Merr.* Tent. 67
Lacerta montana. *Mikan.* Sturm Fauna
Lacerta agilis. *Pennant, Jenyns, Berksnh.*, &c.
Lacerta ædura. *Sheppard.* Linn. Trans. vii., 49
Lacerta Schriebersiana. , *Edw.* Ann. S.N. xvi., 83
Lacerta chrysogaster. *Andr.* in Cat.
Lacerta unicolor. *Kuhl?* Beitr. 121.
Lacerta praticola. *Evers.* N.M. Mosq. iii., 347
Zootoca Jacquini. *Cuct.* Mag. Zool. t. 9
Zootoca Guerini. *Cuct.* Mag. Zool. 9
Zootoca muralis. *Gray.* Ann. N. Hist. 279
Common Lizard. *Jenyns.* Man., p. 293
Scaly Lizard. *Penn.* Biit. Zool. iii., p. 21
Viviparous Lizard. *Bell.* Br. Rep., p. 34

Nimble Lizard. *Bell.* Br. Rept. 34

Harriman. (*Shropshire.*)

4. LACERTA AGILIS. *L.* The Sand Lizard. *P.* 27

    Lacerta agilis. *Linn.* Faun. Suec., 284

    Lacerta Europœa. *Pallas.* Zool. Russ., 29

    Lacerta vulgaris. *Müller.* Z. Dan.

    Lacerta stirpium. *Daud.* Rept. iii. 155

    Lacerta Laurenti. *Daud.* Rept. iii. 227

    Lacerta arenicola. *Daud.* Rept. iii. 230

    Lacerta sepium. *Griff.* A.K. ix. 116

    Lacerta rosea. *Fitz.* in Gray. Cat.

    Lacerta catenata. *Fitz.* in Gray. Cat.

    Seps varius. *Laur.* Syn. 62

    Seps cœrulescens. *Laur.* Syn. 62, 171

    Seps argus. *Laur.* Syn. 61, 161

    Seps ruber. *Laur.* Syn. 62, 169

    Seps stellatus. *Schrank.* Faun. Bo. 117

    Lézard des Souches. *Edw.* Ann. Sci. Nat.

    Lacerta di Linneo. *Bonap.* Faun. Ital.

    Sand Lizard. *Bell.* Brit. Rept. 18

5. LACERTA VIRIDIS. *L.* The Green Lizard. *P.* 33

    Lacerta varius. *Edw.* Ann. Sci. Nat. xvi., 64

    Lacerta chloronata. *Rafin.* fide Gray.

    Lacerta serpa. *Rafin.* fide Gray.

    Lacerta elegans. *Andr.*

    Lacerta smaragdina. *Schinz.* Rept. 99

    Lacerta bistriata. *Schinz.* Rept. 99

    Lacerta bilineata. *Daud.* Rept. iii., 152

    Seps terrestris. *Laur.* Syn. 61, 166

    Lézard piqueté. *Edw.* Ann. Sci. Nat., 1829

    Lézard vert. *Duges.* Ann. Sci. Nat., p. 373

6. ANGUIS FRAGILIS. *L.* The Blindworm. *P.* 39

    Anguis eryx. *Linn.* Syst. Nat. i., 392

Anguis lineata. *Laur.* Syn. Rept., 68, 178
Anguis clivica. *Laur.* Daud. Rept. vii., 281
Anguis bicolor. *Risso.* E.M. iii., 89
Anguis incerta. *Krynick.* Bull. Mcsq., 52
Anguis cinereus. *Risso.* E.M. iii., 88
Siguana Ottonis. *Gray.* Griff An. King. ix., 74
Otophis eryx. *Fitz.* Verz.
Cæcilia vulgaris. *Aldrov.* Serp. xi., 243
Cæcilia typheus. *Charl.* Exer., 31
Cæcilia anglica. *Petiver.* Mus. ii.
L'Orvet. *Lacep.* Quad. Ov. ii., 430
Blindworm. *Penn.* Brit. Zool. iii., 36
Slowworm. *Bell.* Brit. Rept., 41
Hazelworm. *Van Lier.*

ORDER III.—OPHIDIANS.

SUB-ORDER I.—COLUBRINES.

FAMILY.—COLUBRIDÆ.

7. TROPIDONOTUS NATRIX. *Boie.* Common Snake. *P.* 46

Natrix torquata. *Ray.* Syn. Quad., 334.
Natrix vulgaris. *Laur.* Spec. Med., 75
Coluber natrix. *Linn.* Syst. Nat. 380
Coluber torquatus. *Lacep.* Quad. ii., 147
Coluber murorum. *Vest.*
Tropidonotus chersoides. *Dum. and Bib.* (part).
Tropidonotus persa. *Eichw.*
Ringed Snake. *Penn.* Brit. Zool. iii., 33
Common Snake. *Bell.* Brit. Rept. 49

FAMILY.—CORONELLIDÆ.

8. CORONELLA LÆVIS. *Boie.* Smooth Snake. *P.* 51

Coronella Austriaca. *Laur.* Syn. t. 5 fig. 1
Coluber Austriacus. *Shaw.* Zool. iii., 515

Coluber lævis. *Lacep.* Quad. Serp. ii., 158
Coluber Dumfrisiensis. *Sow.* Brit. Misc., 5
Coluber Thuringicus. *Bechs.* Ub. Lacep. iii., 181
Coluber ferrugineus. *Sparm.* N.S. Abhand. xvi., t. 7
Zacholus Austriacus. *Wagl.* Syst. Amph., 190
Natrix Dumfrisiensis. *Flem.* Brit. An., 156
Smooth Snake. *Zoologist*, etc.
Dumfries Snake. *Sowerby.* Brit. Misc.

SUB-ORDER II.—VIPERINES.

FAMILY.—VIPERIDÆ.

9. PELIAS BERUS. *Merr.* Viper.   *P.* 67

Coluber Berus. *Linn.* Syst. Nat. i., 377
Coluber chersea. *Linn.?* Syst. Nat., 337
Coluber cæruleus. *Shepp.* Linn. Trans. vii., 56
Vipera. *Ray.* Syn. Quad., 285
Vipera vulgaris. *Latr.* Rept. iii., 812
Vipera communis. *Leach.* Zool. Misc. t. vii.
Viper. *Penn.* Brit. Zool. iii., 26
Red Viper. *Rackett.* Linn. Trans. xii., 349
Blue-bellied Viper. *Shepp.* Linn. Trans. vii., 56
Black Viper. *Leach.* Zool. Misc. iii.
Adder. *Scotland.*
Common Viper. *Shaw.* Gen. Zool. iii., 365
Etther. *Shropshire.*

O

CLASS.—AMPHIBIA or BATRACHIANS.

A. BATRACHIA SALIENTIA (without tails).

10. RANA TEMPORARIA. *Linn.* Common Frog.   *P.* 91

   Rana aquatica. *Ray.* Syn. Quad., 247
   Rana Muta. *Laur.* Syn. Rept., 30
   Rana Scotica. *Bell.* Brit. Rept. i., 102
   Common Frog. *Penn.* Brit. Zool. iii., 9
   Scottish Frog. *Bell* (variety).

11. RANA ESCULENTA. *Gesn.* Edible Frog.   *P.* 102

   Rana viridis. *Linn.* Faun. Suec., 94
   Rana ridibunda. *Pallas.* Iter.
   Rana cachinnans. *Eichw.* Faun. Casp., 126
   Rana palmipes. *Spix.* Test., t. 5, f. 1
   Rana maritima. *Risso.* Europ. Mer. iii., 92
   Rana alpina. *Risso.* Europ. Mer. iii., 93
   Rana calcarata. *Mich.* Isis, 1830, 160
   Rana tigrina. *Eichw.* Faun. Casp., 125
   Rana Hispanica. *Bonap.* Faun. Ital.
   Edible Frog. *Penn.* Brit. Zool. iii., 13
   Green Frog. *Shaw.* Zool. iii., 103
   Whaddon Organs. *Cambridgeshire.*
   Moor Frog. *Germany.*

12. BUFO VULGARIS. *Laur.* Common Toad.   *P.* 113

   Bufo terrestris. *Schw.* Therio Sil., 159
   Bufo cinereus. *Schneid.* Hist. Amph., 185
   Bufo palmarum. *Cuv.* Regne Anim.
   Bufo alpinus. *Schinz.* Faun. Helv., 144
   Rana rubeta. *Gesn.* Hist. Anim. ii., 59
   Rana bufo. *Linn.* Syst. Nat., 354
   Toad. *Penn.* Brit. Zool. iii., 14
   Common Toad. *Shaw.* Zool. iii., 138
   Paddock. *Bell.* Br. Rept., 115

13. BUFO CALAMITA. *Laur.* Natterjack. *P.* 131

Bufo cruciatus. *Schneid.* Hist. Amph., 193
Bufo rubeta. *Flem.* Brit. Anim., 159
Bufo portentosus. *Schinz.* Faun. Helv., 144
Bufo viridis. *Dum. and Bibr.* (in part), p. 681
Rana fœtidissima. *Herm.* Aff. Anim., 260
Rana portentosa. *Blum.* Handb., 213
Rana mephitica. *Shaw.* Gen. Zool., 152
Goldenback. *Surrey.*
Walking Toad. *East Anglia.*
Rœhrling, or Reed-Frog. *Germany.*
Mephitic Toad. *Shaw.* Zool., p. 152
Natterjack. *Penn.* Brit. Zool. iii., 19

B. BATRACHIA GRADIENTA (with tails).

FAMILY 1.—SALAMANDRINES.

14. TRITON CRISTATUS. *Laur.* Great Water Newt. *P.* 141

Lacerta palustris. *Linn.* Syst. Nat. i., 370
Lacerta aquatica. *a. Gmel.* Syst. Nat. i., 1065
Lacerta lacustris. *Blum.* Handbk., 248
Lacerta porosa. *Retz.* Faun. Suec. i., 288
Lacerta palustris. *δ. Gmel.* Syst. Nat. i., 1065
Lacertus aquaticus. *Gesn.* Quad. Ov. 31
Salamandra palustris. *Schneid.* Amph. i., 60
Salamandra cristata. *Daud.* Rept. viii., 233
Salamandra platyura. *Daubent* in Gray Cat.
Salamandra laticauda. *Bonn.* Enc. Meth. t. 2, f. 4
Salamandra platicauda. *Rusc.*, t. 1, f. 3, 4
Salamandra aquatica. *Ray.* Syn. Quad., 273
Salamandra carnifex. *Schneid.* Amph. i., 74
Salamandra pruinata. *Schneid.* Amph. i., 69
Molge palustris. *Merrem.* Syst. Amph., 187
Triton Americanus. *Laur.* Rept. 49
Triton Bibroni. *Bell.* Brit. Rept., p. 140

APPENDIX.

Triton marmoratus. *Bibr.* Proc. Zool. Soc.
Triton carnifex. *Laur.* Rept., 38
Triton asper. *Higgin.*
Triton palustris. *Jenyns.* Man., p. 303
Great Warty Newt.
Great Water Newt. } *Shaw.* Zool. iii., p. 296
Common Warty Newt. *Bell.* Br. Rep., p. 129
Warty Lizard. *Penn.* Br. Zool. iii., p. 23
Warted Newt. *Shaw.* Nat. Misc. viii., pl. 279
Warty Eft *Jenyns.* Man., p. 303
Salamandre crêtée. *Cuv.* Règ. An. ii., p. 116

15. LOPHINUS PUNCTATUS. *Gray.* Common Newt. *P.* 151

Lacerta aquatica. *Linn.* Syst. Nat. i., 370
Lacerta aquatica. β. *Gmel.* Syst. Nat. i., 1106
Lacerta triton. *Retz.* Faun. Suec. i., 288
Lacerta salamandra. ε. *Gmel.* Syst. Nat. i., 1106
Lacerta tæniata mas. *Sturm.* Fauna Germ.
Lacerta maculata. *Sheppard.* Linn. Trans. vii., 53
Lacerta vulgaris. *Linn.* Syst. Nat. i., 370
Salamandra palmata. *Schneid.* Amph. i., 72
Salamandra tæniata. *Bechs.* in Schneid. i., 58
Salamandra exigua. *Rusc.* t. 1, f. 1, 2
Salamandra elegans. *Daud.* Rept. viii., 255
Salamandra punctata. *Daud.* Rept. viii., 267
Salamandra palustris. *Bonelli* in Gray Cat.
Salamandra palmipes. *Latr.* Rept. ii., 240
Salamandra abdominalis. *Daud.* Rept. viii., 250
Molge punctata. *Merrem.* Syst. Amph., 186
Molge palmata. *Merrem.* Tent., 186
Molge cinerea. *Merrem.* Tent., 183
Triton parisinus. *Laur.* Rept., 40
Triton punctatus. *Bonap.* Faun. Ital.
Triton lobatus. *Otth.* Techudi Batrach., 95
Triton aquaticus. *Flem.* Br. Anim., 158

Triton palustris. *Laur.* Spec. Med., p. 39
Triton lævis. *Higg.*
Triton exiguus. *Laur.* (*young*). Rept., 41
Lissotriton punctatus. *Bell.* Brit. Rept., 142 .
Lissotriton palmatus. *Bell* (*aged*). Brit. Rept.
Brown Lizard. *Pennant.* Brit. Zool. iii., 23
Common Newt. *Shaw.* Gen. Zool. iii., 298
Smooth Newt. *Bell.* Brit. Rept. 143
Man-eater. *Ireland.*
Man-keeper. *Ireland.*
Dry Ask. *Ireland.*
Dark Lewkers. *Ireland.*
Small Newt. *Bell.* Brit. Rept., 143
Eft, or Evet. *Bell.* Brit. Rept., 143
Asgal. *Shropshire.*

16. LOPHINUS PALMATUS. *Dum. and Bibr.* Palmate Newt.
P. 160
Salamandra exigua. *Laur.* (*young*). Rept. ♀
Salamandra abdominalis. *Daud.* Rept. ♀
Salamandra cincta. *Daud.* Rept. viii., 259
Salamandra palmata. *Cuv.* (not *Schneid.*) R.A., ii.
Triton palmatus. *Bonap.* Fauna Ital.
Triton minor. *Higg.*
Triton palmipes. *Deby.* Zool., p. 2233
Lissotriton palmipes. *Bell* (2nd ed.), p. 154
Palmated Smooth Newt. *Bell.* Brit. Rept., 154

17. OMMATOTRITON VITTATUS. *Gray.* Gray's Banded Newt.
P. 168
Salamandra vittata. *Gray.* Brit. Mus.
Molge vittatus. *Gray.* Brit. Mus., 1830
Triton vittatus. *Jenyns.* Manual, p. 305
Lissotriton palmipes, var. *Bell.* Brit. Rept., p. 152
Striped Eft. *Jenyns.* Man., 305.

INDEX.

www.ingramcontent.com/pod-product-compliance
Lightning Source LLC
Chambersburg PA
CBHW021657210326

41599CB00013B/1452